FUTURES IN THE MAKING

FUTURES IN THE MAKING

SPECULATIVE ART FOR A BURNING WORLD

THEO REEVES-EVISON

THE MIT PRESS
CAMBRIDGE, MASSACHUSETTS
LONDON, ENGLAND

The MIT Press
Massachusetts Institute of Technology
77 Massachusetts Avenue
Cambridge, MA 02139
mitpress.mit.edu

The MIT Press would like to thank the anonymous peer reviewers who provided comments on drafts of this book. The generous work of academic experts is essential for establishing the authority and quality of our publications. We acknowledge with gratitude the contributions of these otherwise uncredited readers.

This book was set in Arnhem Pro and New Frank by New Best-set Typesetters Ltd.
Printed and bound in the United States of America.

Library of Congress Cataloging-in-Publication Data

Names: Reeves-Evison, Theo author
Title: Futures in the making : speculative art for a burning world / Theo Reeves-Evison.
Description: Cambridge, Massachusetts : The MIT Press, [2026] | Includes bibliographical references and index.
Identifiers: LCCN 2026004015 (print) | LCCN 2026004016 (ebook) | ISBN 9780262056120 paperback | ISBN 9780262056137 epub | ISBN 9780262056144 pdf
Subjects: LCSH: Human ecology in art | Future, The
Classification: LCC N8217.E28 R44 2026 (print) | LCC N8217.E28 (ebook)
LC record available at https://lccn.loc.gov/2026004015
LC ebook record available at https://lccn.loc.gov/2026004016

10 9 8 7 6 5 4 3 2 1

EU Authorised Representative: Easy Access System Europe, Mustamäe tee 50, 10621 Tallinn, Estonia | Email: gpsr.requests@easproject.com

For Rowan

CONTENTS

INTRODUCTION

PATCHWORK FUTURES

How are the futures of climate change and biodiversity loss imagined? This is the deceptively simple question that opens this book. Look beyond terrifying predictions of scorched or flooded worlds, mass extinction events, conflicts over dwindling resources, and people robbed of their lives or livelihoods. These events are already unfolding in many parts of the world, and their urgency cannot be underestimated. But they are not answers to a "how" question; they are answers to a "what" question. There are now many thousands of articles and reports that set out the scientific basis of what can be expected over the coming decades, as well as countless books that help package this research for a general audience. Instead of fueling this engine of anticipation and offering another take on what environmental futures may entail, this book looks at some of the creative means by which such futures are conceptualized and produced. Its primary aim is not to find out where the world is heading but to map some of the bridges and pathways that have been used to get there.

The question of how futures are imagined leads very quickly to a consideration of the role of images and visual culture. From the apocalyptic scenarios that fill cinema screens to the data visualizations that cascade from research organizations to news reports, images that speak to the futures of climate change and biodiversity loss have increased in number and exposure over the past few decades. The influence of these visual practices is not, however, solely exerted through the representations they generate. While these may attract the most attention, it is equally important to direct our gaze to the unobtrusive tools and techniques that inform how such images link pasts, presents, and futures—an attention that, in turn, engenders agency. *Futures in the Making* traces the genealogies of how some of these techniques have come to shape the story of environmental change as it extends into multiple futures. In doing so, it offers a partial history of some of these practices, identifying some modes of travel to environmental futures and suggesting ways such modes can be expanded through creative practice today.

Before turning to art and visual culture, let us first cast our attention on what is arguably the keystone scientific organization for gathering and communicating climate knowledge today: the Intergovernmental Panel on Climate Change (IPCC). The IPCC's mode of communication is typically unidirectional, with each report triggering a cascade of headlines, interviews, and social media activity around the world. Projections, models, and scenarios fill the news; windows for action on climate change open and close; and more progressive politicians ramp up the rhetoric on putting the interests of younger generations before their own. But how are the ominous climate futures that circulate in this swirl of media attention generated? What are the tools and techniques that shape how such visions are produced and articulated? And what effects do these futures have on our pathways to action in the present?

A diagram entitled "Characterisations of the Future" (figure I.1) provides a good starting point for thinking about the various methods the IPCC and broader scientific community use to anticipate the effects of climate change.[1] A patchwork of overlapping rectangles provides the background for labels such as "Projections," "Sensitivity analysis," and "Scenarios & Storylines." Anticipating the effects of climate change, the diagram suggests, is a fundamentally heterogeneous activity, guided by a variety of methods that help actors navigate through the unfolding environmental crisis in different ways.

Like any diagram in a publication of this nature, the aim is to present information clearly, with the stated intention to offer "a consistent typology" for the purposes of clarifying terminology used elsewhere in the report.[2] Methods are arranged in relation to one another based on their comprehensiveness, plausibility, and whether they operate on a

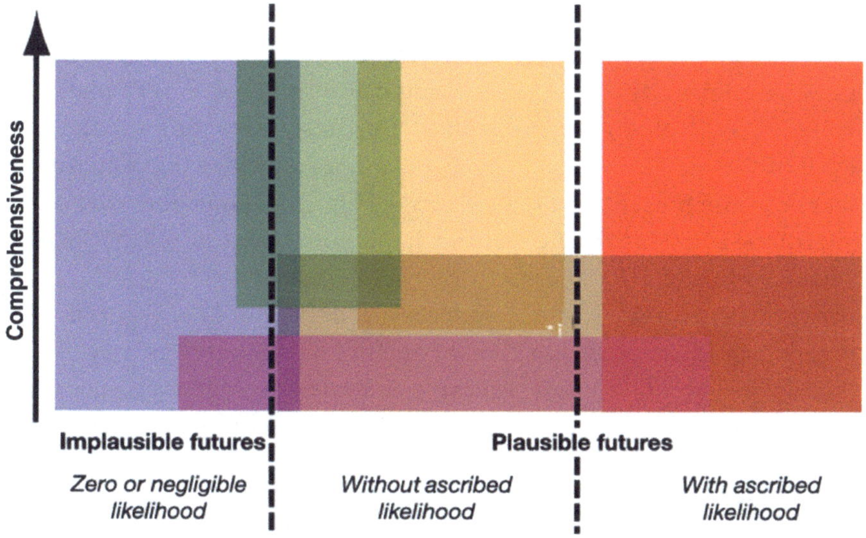

quantitative or qualitative basis. Some boxes invade the space of their neighbors, creating an epistemological moiré pattern that frustrates any attempt to see them in clear distinction from one another. The illusion of simplicity is further eroded when the histories of the various labels on the diagram are further unpacked. "Probabilistic Futures," for example, would be unimaginable without modern approaches to the mathematics of uncertainty, and large-scale projections were only made practicable after the invention of computer modeling.[3] As many have pointed out, the emergence of global circulation models combining oceanic and

FIGURE I.1

IPCC, "Characterisations of the Future." Source: Figure 2.4 in Timothy R. Carter et al., "New Assessment Methods and the Characterisation of Future Conditions," chap. 2 of *Climate Change 2007: Impacts, Adaptation and Vulnerability. Contribution of Working Group II to the Fourth Assessment Report of the Intergovernmental Panel on Climate Change*, ed. Martin Parry et al. (Cambridge University Press, 2007), 145.

atmospheric processes in the 1960s played a particularly important role in increasing understanding of environmental futures; these steadily grew in complexity as they incorporated additional aspects of planetary systems.[4] With enough of this context, the diagram extends into the past: Each of the boxes has a backstory, a distinct genealogy that determines its use in the present and the futures it helps us to imagine. One of the jobs of the IPCC is to orchestrate such methods, to create an ensemble of what we will come to understand over the course of this book as "speculative infrastructures" that help shape a world that does not yet exist.[5] The diagram, in this sense, is a map of such infrastructures as they extend into environmental futures. And yet, as many have observed, by orchestrating speculative practices in this way, the IPCC is adopting not only the role of mapmaker for such futures but also that of "corridor maker," actively reducing the discursive space in which environmental futures can be conceived.[6] Suddenly, the deceptively simple question of how the futures of climate change and biodiversity loss are imagined becomes a lot more complex.

One of the central ambitions of this book is to help widen such corridors of the imagination and take the multiplicity of speculative practices seriously. For while "the future" may be spoken of in the singular—a linguistic norm that already associates it with predictive thinking—the tools that help shape it as an object of experience are numerous and diverse. Within the physical and social sciences there are many well-established methods of characterizing futures, from data-driven practices such as global circulation models to more openly subjective exercises in planning, trend analysis, and "conjecture."[7] Indeed, there has been an increasing reliance on such data-driven practices in recent decades, based on the assumption that better models will automatically lead to more knowledge, which in turn translates into collective shifts in behavior.[8] The role of the arts and visual culture in this process is generally relegated to the sidelines. And yet the arts have generated numerous innovations in how futures are imagined, both independently and in dialogue with scientific anticipatory practices—anticipation, here, being a term that represents a subset of speculative practices that carries associations of preparedness, pragmatism, and of "rendering the future actionable."[9] New futures come into being by virtue of the new tools used to imagine them, and the business-as-usual futures that sometimes seem most likely are often revealed to be the product of what Barbara Adam calls "habits of mind," restricted by their lack of plasticity.[10] The chapters that comprise *Futures in the Making* are an attempt to navigate through this broader cultural landscape of speculative practices, from those that encourage multi-sensory and embodied responses to multiple futures to practical enactments that cultivate post-growth cultures in the present. The result is a patchwork of tools, techniques, and methods, both formal and informal, that overspill the boundaries of any IPCC diagram, set against a background in which it has become impossible to think seriously about futures without simultaneously thinking about the catastrophic and uneven effects of ecological breakdown and climate change.

MAKING FUTURES WITH CREATIVE PRACTICE

It is not uncommon today to find artists experimenting with forms of science fiction, world-making, and futurisms of various kinds, often impelled by environmental crises operating at a variety of scales. In parallel to this, and after several decades in which a perceived lack of forward momentum in creative practice was a common diagnosis, many cultural theorists have also turned their attention to futures once again, as will be discussed later in this introduction. To understand some of this recent attention on futures and futurisms, there is much to be gained by an engagement with the past, and in the case of environmental futures, the past of the mid-twentieth century, specifically. While this was a period in which nuclear disaster was seen to pose more of an existential threat to humanity than climate change and biodiversity loss, it was also a time in which many of the techniques currently used to imagine environmental futures were first developed.[11] Needless to say, futures had been the subject of inquiry for diverse knowledge traditions millennia before this, as Wendy Schultz acknowledges in her compressed history; however, it is in the postwar period that the story of modern futures studies begins, characterized by methodological experimentation and competing political interests.[12]

Against the backdrop of a bipolar world, figures such as Herman Kahn and Alvin Toffler became international celebrities with bleak reflections on societal collapse and nuclear holocaust, while on the other side of the Iron Curtain, centralized five-year plans belied a more complex range of future practices.[13] It was during this period that the word "futurist" was consolidated as the name of a recognized profession in many Western countries, with all who claimed the title listed in a *Futurist Directory*.[14] While Kahn engaged in pioneering futures work from deep within the US Cold War military establishment, another side of futures studies was emerging from the global environmental and peace movements of the 1960s and '70s, exemplified in organizations such as the Club of Rome and the World Futures Studies Federation. For Ahlqvist and Rhisiart, these pathways represented two competing trajectories within futures studies, with the "emancipatory" tradition steadily coming to be overshadowed by a utilitarian approach placed in the services of military, policy, and business—the latter epitomized in the rise of "corporate foresight."[15]

It was in this context that many of the methods, concepts, and assumptions of modern-day futures studies were established, which, for much of its existence, provided a big enough tent to contain both utilitarian and emancipatory impulses. And yet according to Hyeonju Son, this steadily gave way to a fragmentation in the 1990s, with the goal of much futures work being the success of specific organizations and projects rather than the development of shared civic or planetary agendas.[16] The fragmentation of the field in the 1990s is also reflected in the number of academic initiatives engaged in studying futures. Here it is worth briefly summarizing three prevailing tendencies before considering the place of creative practice in relationship to futures work, and in this book as a whole.

Of all established disciplines, it is perhaps sociology that has had the longest relationship with futures work, beginning with figures such as Wendell Bell attempting to found a sociology of the future in the 1960s and '70s.[17] While some of this work was influential in interdisciplinary arenas, it was ultimately unsuccessful in carving out a dedicated niche for itself within sociology (as Bell himself noted in the 1990s).[18] Aside from the influential work of Barbara Adam, whose concepts crop up at various points in this book, discussions of futures over the past few decades have mainly entered sociology through the side door of science and technology studies (STS), adding to a growing literature on what has also come to be known as the "sociology of expectations."[19] This literature places particular emphasis on the role of representations, visions, and promises in shaping scientific and technological change, a subject that will be picked up later in the introduction and in subsequent chapters.[20]

In recent years, sociology has been joined by the relatively new arrival of anticipation studies, which constitutes a much broader field of study for its proponents. Anticipation is positioned here as a widespread phenomenon, "embedded in all kinds of living beings," and with a semantic range that exceeds its use in this book.[21] As a result of this range, anticipation studies necessarily straddles the natural and social sciences, finding examples of anticipatory activity in various aspects of behavior and cognition, social processes, and even non-living systems.[22] It also has a practical dimension, with key figures such as Riel Miller embedding some of its approaches within organizations such as UNESCO, where he served as Head of Foresight and Futures Literacy—the latter understood as a way of encouraging greater reflection on the underlying assumptions that shape how futures are imagined.[23]

This links to a third field, or more accurately subfield, which is where this book sits most comfortably: critical futures studies (CFS). Although broadly part of the emancipatory tradition identified by Ahlqvist, Rhisiart, and others, it is productive to see CFS as a counterweight to the development and dominance of "foresight" since the 1990s. Foresight has a pragmatic orientation in which specific methods are put to work for corporate and policy gains, often with a focus on science, technology, and innovation. For Sardar, it should therefore fundamentally be understood as a tool of management, "with a clear emphasis on winning over others, instead of exploring and developing creative, novel and inclusive solutions."[24] While the picture is more complicated when it comes to civic or national applications, it would be accurate to say that foresight generally fits comfortably within market-oriented societies, often circumventing issues of value, morality, inequality, and power.[25] It is nevertheless underpinned by a robust set of research methods, often at the expense of wider theoretical development.[26]

By contrast, a key characteristic of CFS is that it looks outside of the confines of corporate or policy worlds, building on foundations laid by environmental and peace researchers in the 1960s and '70s, such as

Johan Galtung, Robert Jungk, and Elise Boulding.[27] More populist writers, such as Alvin and Heidi Toffler, were also critical of what they saw as the concentration of futures expertise, instead proposing networks of "imaginetic centres" and "social future assemblies" that would involve diverse groups in acts of speculation.[28] Today, while participation in futures work has widened a great deal, what was criticized in the 1970s as "ivory tower" futurism is still alive and well.[29] Tech firms and consultants market insights about futures gained by proprietary means, policy think tanks engage in horizon-scanning initiatives with few citizens understanding what this entails, and data modeling has developed into a specialized pursuit only available to those with the right technical know-how, software, and computing power.[30] Despite genuine and ongoing attempts on the part of some governments and NGOs to broaden participation in futures work, from the perspective of more progressive practitioners working in the field today, futures literacy among the general public remains low.[31]

Closely aligned to this democratizing tendency within CFS is the ambition to repoliticize futures work, which involves exposing the values and "embedded cultural commitments" within futures studies (a reflexive charge led by significant feminist contributions to the field), as well as the ways in which futures practices and discourses reinforce existing socioeconomic formations.[32] Claims to technocratic expertise within corporate foresight and policy-making supply a steady stream of fodder for this kind of critical work, but there is also a need for CFS to turn this critique back upon itself, interrogating its own sociopolitical commitments.[33] For Sardar, doubt plays an important role, casting a skeptical gaze on "simple, one-dimensional solutions to wicked problems as well as . . . dominant ideas, projections, predictions, forecasts and notions of truth to ensure that the future is not foreclosed and colonised by a single culture."[34] In its appeal to a broad range of cultural practices, this book participates in a shift within futures studies to expand the terms and terrain upon which its work takes place.

This is the first of several arguments this book makes about art and creative practice: that it has the capacity to expand the frame of what constitutes a valid form of future-making.[35] When it comes to environmental futures, scientific cultures of prediction and modeling have delivered invaluable insights about the state of the planet, but if their power comes at the expense of other speculative practices within the arts and humanities, then this only serves to reduce the cultural bandwidth of possible futures.[36] Sometimes artists draw direct inspiration from futures studies and the environmental sciences, as discussed in chapter 1, which deals with the historical and current use of scenario planning, and chapter 3, which considers the epistemological grounding of geoengineering. At other times, artists have worked in direct opposition to the perceived conservatism of what Peter Galison, in broad brush strokes, has called "state science fiction."[37] Another, no doubt larger, group of artists makes future-oriented work independent of mainstream futures research. By bringing this last group into contact with some of the debates and discussions within more scholarly forums, the aim is to cross-fertilize otherwise disconnected areas of futures work and make the points of commonality and divergence more apparent. This is the first book of its kind to devote serious attention to this task, interweaving discussions of futures research and creative practice with cultural and environmental theory.

A significant portion of existing writing within CFS engages in critical and self-reflexive work by focusing on discourse, as was common in much critical scholarship of the 1990s and 2000s. The work of Slaughter and Inayatullah, for example, routinely places emphasis on "the re-negotiation of meanings," which has the capacity to expose the present to be a "fragile . . . victory of one particular discourse, way of knowing, over another."[38] Without diminishing the power of discourse, stories, fiction, and other reality-altering social technologies, it is important to also consider the material dimensions of how futures are imagined, produced, and, in some cases, resisted.[39] For while certain realities are indeed fragile, others are stubbornly entrenched, locked in to processes that persist despite shifts in narrative or discourse. Our present words and deeds send ripples through language that lap at the shores of tomorrow, but they also congeal into socio-material relations that endure over time.[40] As discussed in the next section, the concept of infrastructure will be crucial in thinking these two dimensions of discourse and materiality in combination.

Art and visual culture, while frequently analyzed on a discursive plane, also have an important part to play in mediating matter and meaning. When entangled with scientific, corporate, and governmental organizations that produce knowledge about futures, creative practice can help bring this knowledge to greater visibility in a range of ways. This is a second argument for how art and visual culture can be meaningful in the context of environmental futures: not simply as cultural artifacts that can pry open the histories and practices of futures thinking but also as instruments that can concretize tools and techniques that might otherwise be difficult to identify because they seem intangible. Futures work and the broader encompassing field of speculative practices could be considered to contribute forms of knowledge infrastructure. As in Trevor Paglen's photographs of the hidden undersea internet cables that keep digital networks connected (figure I.2), art can bring these knowledge infrastructures to the representational surface of the work, providing momentary glimpses of vast social, technological, and epistemological systems, and, in doing so, unveil some of the discreet power they exert.[41] Moreover, through diverse aesthetic strategies, creative practice can transform environmental futures from objects of thought and rational planning into embodied sensations, or what we could call, repurposing a term from Ruth Wilson Gilmore, "infrastructures of feeling."[42]

This passage to visibility is entirely in keeping with the visual origins of the word "speculation," which stems from the Latin *specere*, "to look or behold." This meaning persisted in English until the mid-nineteenth century, when speculation was used to describe the faculty of vision, and especially "comprehending vision," underpinned by forms of intelligence.[43] While this meaning might be obsolete today, speculation is nevertheless still bound up in practices of seeing and visualization in profound ways (one of the reasons it is used as an umbrella term to encompass the diverse futures practices this book discusses).[44] This is the third proposition this book makes about how art and visual culture can be approached: as tools that facilitate speculation—in both senses of the word. For example, in chapter 2, "Prospecting," practices of surveying and mapping a territory from above are shown to play a vital albeit complex role in what Macarena Gómez-Barris calls the "extractive view."[45] These visual tools are utilized not only in the search for future reserves of fossil fuels and minerals but also in other forms of prospecting, to the extent that it would be difficult to imagine how the practice would function without them. Further filtering visual cultures of extraction through works of art facilitates a "view on the view," so that the scopic regimes that condition how

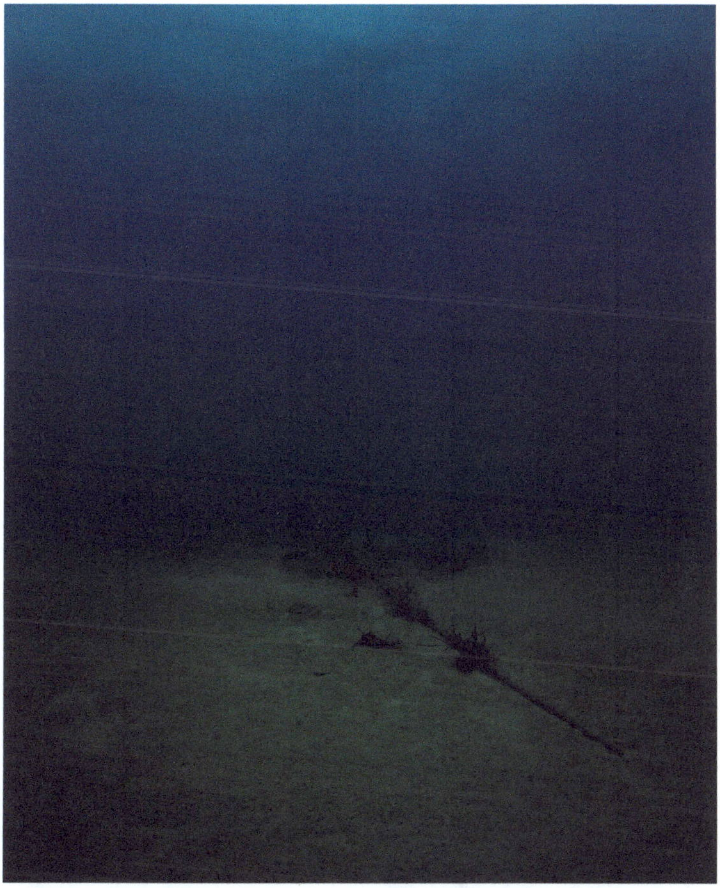

FIGURE I.2

Trevor Paglen, *Bahamas Internet Cable System (BICS-1), NSA/GCHQ-Tapped Undersea Cable, Atlantic Ocean*, 2015, C-print, 60 × 48 in. Courtesy of the artist.

the futures of resource extraction are conceptualized become objects of reflection themselves, rather than unacknowledged "operational images" that largely pass unnoticed.[46]

Understanding how visual cultures and techniques have helped capitalism to conceive of the world as a storehouse of resources in the past has a key role to play in helping to counter the continued hegemony of these framing devices in presents and futures. Art that does this is largely critical, complementing a materialist CFS by scrutinizing the visioning techniques of extractivism, colonialism, and technocapitalist solutions to climate change. As theorists of ecology and visual culture such as T. J. Demos attest, however, there is a more propositional role that art can play in relation to environmental futures. As well as providing a lens through which to look at the world, or a "view on the view," contemporary art can also function as "an intense site of assorted world-making activities."[47] The futures that emerge from such speculative cultural practices are not the visions of heroic artists working in isolation but rather, for Demos, are "situated within or allied with communities of action and social movements, and always, most ambitiously, provoking and furthering an entanglement of insights, perceptions, affects."[48] This is ecology not only as subject matter but also as *modus operandi*—inspired by the "three ecologies" of Félix Guattari—in which the world-making capacity of art is extended far beyond the typical realm of "green politics," narrowly understood. In this approach, art becomes enmeshed within social and mental ecologies too—in other words, precisely the ecologies that are elided by much scientific anticipatory practice.[49]

There are examples of the generative, world-making capacity of art woven through this book, often in cultural practices that combine methodological innovation with clearly articulated critical positions. Encouraging such methodological innovation in art and beyond is a key ambition of this book, shifting the focus away from the "what" of environmental futures and toward the "techniques of *futuring*" that often escape notice.[50] In the final chapter—which seeks to theorize the concept and practice of enacting—the register becomes most obviously propositional, as the discussion turns toward environmental futures that don't take the benevolence of endless economic growth as an unquestioned assumption. These examples are no less practical for the fact that they are speculative; indeed, they are animated by the urgency of the present and frequently at odds with the gestural politics of much of the contemporary art world. Forms of symmetrical reciprocity and solidarity economies may seem like leftovers of outmoded economic systems, but they can equally serve as the basis for the kinds of post-growth futures that the climate crisis demands. Emphasizing such practical speculations is also one way of redirecting attention away from the representation of environmental futures in art and visual culture and toward the concrete processes that bring livable worlds into being.

SPECULATIVE INFRASTRUCTURES

As mentioned, one of the propositions this book makes is that the various tools, techniques, and methods that bridge our relationship to futures represent a kind of infrastructure. While the opening pages of this book have already conveyed some of the senses in which this term is being deployed, it is worth digging a bit deeper here to think through some of the subtleties of what it might mean in the context of environmental knowledge and creative practice.

As the STS pioneer Susan Leigh Star observed in the 1990s, infrastructure is commonly understood as a substrate, "sunk into other structures," or the base layer upon which something else operates.[51] Trains run on the infrastructure of tracks; cars run on the infrastructure of roads; the internet runs on the infrastructure of servers, cables, and routers. Such a conception has its advantages, insofar as it highlights the hidden nature of much infrastructure, and this invisibility goes some way to explain why it is often overlooked.

And yet this characterization can also privilege everyday physical infrastructure at the expense of knowledge infrastructure, which may have less obviously material dimensions than the cables, pipes, and tracks that constitute the physical backdrop to much of human life. Trains, after all, do not only run on the infrastructure of tracks; they also rely on timetabling schedules, signaling protocols, and numerous other systems and processes. Taking climate knowledge infrastructure as his key example, Paul N. Edwards speaks of the material network of things such as computer systems, technical instruments, and human beings, alongside intangible aspects of infrastructure such as assessment processes, specialized vocabularies, and conventions on intellectual property.[52] Where early discussions of infrastructure in the social sciences took physical systems as their starting point, there is now a sizeable literature, both within STS and beyond, that lays equal stress on the immaterial dimensions of how infrastructure comes to shape our world. As an example of this, consider again the strategies that organizations such as the IPCC use to maintain trust, the desire for which influences everything from the grammar and syntax used in reports to the sequence in which they are released to the public.[53] Without the trust the organization commands, the climate futures it generates might be considered "wild" speculation, unmoored from the knowledge infrastructures that support their claims. As discussions of infrastructure spread beyond STS into fields where the category is used to understand an ever-broader range of phenomena, the term's semantic range inevitably expands in the process. In the last decade or so, "infrastructuralism," "infrastructural critique" and the "infrastructural turn" have shaped discussions in diverse disciplines in the humanities, such as literature, history, and philosophy, leading to a situation in which it becomes hard to find socio-cultural phenomena that could not be considered infrastructure of one sort or another.[54] On the other hand, there are also instances when stretching the semantic limits of the term in this way can be analytically purposeful, for example, when Tlingit scholar Anne Spice recounts the words of Freda Huson, spokesperson for an anti-pipeline encampment in the Unist'ot'en territory in northern British Columbia, who counters the framing of oil pipelines as "critical infrastructure" for the state by arguing that blackberry bushes are critical infrastructure for Indigenous communities in the region.[55]

In an effort to find a balance between metaphor and materiality and limit semantic overreach, it is useful to reflect on the borders that separate "wild" speculation from its disciplined relation, because these are also the lines at which knowledge infrastructures meet their outside.[56] Take the label "Scenarios & Storylines" from the diagram above, and transport this to another context such as screenwriting, in the sense that "storylines" might thread their way through a movie or television series. Suddenly the scientific authority of a storyline to mediate our relationship to futures is undermined—in spite of the fact that scenario planning as a way of relating to futures has links to Hollywood and, stretching even further back, to the Commedia dell'arte, as we shall see in chapter 1.[57] When scenario planning is formalized into a distinct method of futures studies, this genealogy is erased and the meaning of the word "scenario" is standardized, as (for the most part) it becomes cocooned within the worlds of corporate, scientific, and public-policy research. In other words, the process of forming the future into scenarios is transformed into an infrastructure, which in turn is nested within larger infrastructures and socio-technical systems.

The fuzzy edges of what now appear as formal techniques of planning and prediction are addressed in Jamie L. Pietruska's captivating history of forecasting in late nineteenth- and early twentieth-century America. Pietruska shows how words such as "prediction," "prophecy," and "forecast" were used interchangeably until their divergence toward the beginning of this period.[58] In an account wherein fortune tellers jostle with meteorologists and agricultural statisticians, *Looking Forward* attends to a wider "culture of prediction" that eventually gave rise to systematized forecasts underpinned by knowledge infrastructures and professional authority. In chapter 1, my own account picks up a similar thread several decades after Pietruska's ends, in a period when awareness of climate change had just begun to register in a select number of scientific research areas. The emphasis here is not on how this interest grew into the more contemporary scientific techniques and processes for anticipating climate change, since this topic has already been addressed in some detail elsewhere.[59] The focus, rather, is on sketching an earlier and broader context in which futures work (sometimes passing under the name of futurology) was accorded a key strategic role in the military-industrial apparatuses of the Cold War.[60] Think tanks and policy advisors plotted the hypothetical moves and countermoves of a possible World War III, and the state indulged in the production of a particularly dry subgenre of science fiction in their future-oriented policy reports.[61] In an attempt to differentiate these planning processes from wild speculation, Cold War futures work was framed as a body of scientific knowledge involving "social technologies" and "predictive instrumentalities," underpinned by the burgeoning fields of cybernetics, systems analysis, and decision theory.[62]

The way in which speculations are framed to varying degrees as scientific, objective, and "policy relevant" highlights the work required to maintain the knowledge infrastructures that underpin them. As with physical networks of roads, pipes, or servers, decay creeps into all knowledge infrastructures eventually. Instead of the shovels of tarmac, rolls of duct tape, or hardware patches, maintenance work on knowledge infrastructures might involve bringing divergent processes into alignment, policing infringements of intellectual property laws, or responding to external challenges to the validity of the knowledge generated. For an example of this latter form of maintenance, take the highly influential *Limits to Growth* report published by the Club of Rome in 1972. Underpinned by a systems dynamics computer model called World3, developed at MIT to simulate future trends, the report represented a key moment in the use of computer modeling. For all the headlines and international book sales, *Limits to Growth* also had many critics, both within the scientific community and among those with an ideological opposition to its core message that increasing human resource use was unsustainable.[63] It is partly by virtue of the maintenance work carried out on the knowledge infrastructure underwriting the report that world models have started to be treated as viable tools for anticipating environmental futures again today.[64]

Looking at speculative and anticipatory practices as themselves involving infrastructures—both material and discursive—has the potential to shed light on the forms of power embedded within them. As STS scholars routinely point out, the crucial role of infrastructures in our everyday lives is all too often taken for granted. In the words of Brian Larkin, "we often see computers not cables, light not electricity, taps and water but not pipes and sewers," and, we might add, we often see stand-alone projections, models, and scenarios but not the epistemic cultures that produce them.[65] This, again, is where art and visual culture can intervene, making the invisible visible, sometimes interrupting the smooth functioning of infrastructures in creative ways. This can amount to a form of "unblackboxing," whereby the power that infrastructures exercise is thrown into sharper relief.[66]

What distinguishes this form of power from more overt displays of authority is the capacity of infrastructures to coordinate activities that might lie outside of their direct purviews. An early theorization of infrastructural power was made in the 1980s by the sociologist Michael Mann.[67] Contrasting infrastructural power to "despotic power," which operates "without routine [or] institutionalised negotiation with civil society," Mann spoke of the ability of capitalist democracies to exercise power by penetrating and coordinating the activities of civil society through its own institutional forms.[68] An expanded version of this argument has been made by the architectural theorist Keller Easterling, who sees infrastructural power as a medium of "*extrastatecraft*—a portmanteau describing the often undisclosed activities outside of, in addition to, and sometimes even in partnership with statecraft."[69] In the diverse examples Easterling cites, from the proliferation of special economic zones to the shared standards established by the International Organization for Standardization, "the crucial information about a political bearing is often found not in declaration but in disposition—in an immanent activity and organization."[70]

Such power can sometimes be exerted over long periods of time too. A useful concept that surfaces at various points in this book is that of "lock-in," which describes the way in which certain infrastructures organize activities by cultivating path-dependencies that make alternatives appear too costly to contemplate.[71] Lock-in is an important concept in discussions of infrastructure, but it is also a useful bridge between the discursive and material aspects of futures research. As Tutton puts it, "every materialized future leaves traces that cannot be undone . . . these traces can become path dependencies that 'lock in' certain options and become irreversible because resources used in one way cannot be used again."[72] As the case of fossil fuel infrastructure clearly demonstrates, the entrenchment of certain socio-technical systems often extends far beyond their apparent sphere of influence or intended use, sometimes enduring for years, decades, and even centuries.

The kinds of knowledge infrastructure that are used to anticipate environmental futures may not always be so stubbornly entrenched, but they too exercise a power that is nevertheless inconspicuous. Orchestrating activities in the background, these knowledge infrastructures are seldom the subject of media attention themselves. In this respect, the IPCC diagram is a rare peek under the bonnet of the tools and methods the organization uses to apprehend futures as objects of experience in the present. Appearing deep within a thousand-page report, it is nevertheless a moment of exposure that few non-specialist readers are likely to have experienced. Even those that do read IPCC reports cover to cover are presented with a somewhat flattened representation of the infrastructures of speculation that underpin them. For example, there is little detail on how "likelihood" is ascribed to "probabilistic futures," nor mention of important precedents for scenario planning within the fossil fuel industry. This lack of detail is to be expected from a diagram, but it is nevertheless detail that would provide clues as to the disposition of these infrastructures. This can be highly consequential for the kinds of futures such infrastructures create, opening and closing climate corridors beyond the current impasses of extractive capitalism.

GREAT EXPECTATION MACHINES, OR FUTURE AS A VERB

It is difficult to use the word "speculation" today without simultaneously conjuring the world of speculative finance, full as it is with increasingly exotic instruments for managing—and creating—the risks and uncertainties associated with environmental change. This is another reason this book diverges from other scholarship that reserves the word "speculation" for exploratory futures work unburdened by questions of likelihood, because this usage excludes a great deal of what the word encompasses.[73] In financial speculation, futures are bounded objects, the subject of derivative contracts that can be bought and sold at a mutually agreed price. But as sociologists of finance have shown, the inexorable rise of such derivative contracts is intimately bound up with the theories that were supposed to passively explain them. As Donald Mackenzie argues, the role of finance theory and the mathematical models upon which it is based is not that of a "camera" that merely records empirical facts (if, indeed, we can reduce photography to such a limited function), but of an "engine" that has a part to play in their production.[74] Just as financial theories and models are an active part of economic life, so too speculation is never simply a passive act of observation and recording, but always in some way productive of that which it observes.

This aspect of futures work has not been lost on successive waves of scholars in both sociology and anticipation studies. As representatives of the former, Borup and colleagues identify how expectations in science and technology have material effects, for example, by brokering relationships, setting out shared agendas, attracting investment, and creating obligations through promissory statements that are later used to hold scientists, technologists, and the organizations in which they work to account. Such statements, they show, take on substance, "becoming materially embedded in structures, routines, systems, matters."[75] Anticipation studies, for its part, broadens this discussion to account for the relationship between "looking into the future and taking account of the consequences of that looking, i.e. as the impact of an anticipation on current behavior."[76]

The same could be said of the broader range of infrastructures used to generate representations of futures. Regardless of what some of their proponents might claim, tools such as forecasting, scenario planning, and horizon scanning are active coauthors of the futures they envision, constituting techniques of futuring (and, indeed, "defuturing," insofar as they can bring about path-dependencies).[77] As historian of futurology and futures studies Jenny Andersson puts it, prediction "should be understood as a technology of future making and world crafting, a social and political technology in the Foucauldian sense."[78] This holds true for many of the other techniques developed in the heyday of futures research, which may not have always had impact in exactly the way futurists imagined, but nevertheless became key tools in the boxes of governance and business management, and from there continue to shape the world in enduring ways.

Looking at these tools closely shows that futures are not stable objects that exist independently of the practices and processes that are used to cultivate them. The speculative methods discussed are not simply "characterisations of the future," as the title of the IPCC diagram suggests; they are a series of actions upon a range of futures. The power of speculation is, in this sense, considered to be in part performative, acting as an engine rather than a camera, or what the former investment banker Tony Golding calls a "great expectation machine" that channels energy and attention along specific pathways.[79] Another useful way of thinking about the performativity of future-making practices is by drawing on dramaturgy, not only because this speaks to the origins of scenario planning, as will be elaborated in more detail in chapter 1, but because it opens up a conceptual vocabulary of scripting, staging, and rehearsal that extends to other futures techniques, in particular the idea of enacting possible futures, as discussed in chapter 4.[80] Focusing on the spaces and structures that underpin performances (scripts, stages, sets, audiences, and so on), allows us to avoid binary distinctions between "open" and "closed" futures, and again encourages reflection on how socio-material conditions impose limits on what imaginaries can and cannot be articulated in the theatre of the popular imaginary.

Emphasizing the performative dimensions of future-making practices is not, however, the same as arguing that all futures are constructed equally, and therefore equally likely to happen. The "Shared Socio-Economic Pathways" set out in IPCC reports, underpinned as they are by scenario planning and climate modeling, are, needless to say, significantly more robust than the futures imagined by conspiracy theories circulated on social media or the coordinated misinformation campaigns spread by fossil fuel companies. The idea that the climate is changing due to the activities of *some* humans is now an established fact.[81] And yet the speculative infrastructures that have helped produce this established fact have a history, a disposition, and a community of users who deploy them in certain ways and not in others. For example, the IPCC has been criticized for its relatively conservative approach to scenario planning, for the most part choosing to ignore climate futures that are considered too utopian or dystopian, thereby reinforcing the dominant ideology whereby the basic operations of capitalism extend into futures unaffected by a catastrophe they have created.[82]

Without diminishing the importance of the futures such scientific works generate, the bracketing of more progressive futures ensures their invisibility and consequently diminishes their capacity to catalyze action in the present. It can therefore be said that this use of scenario planning acts upon possible futures, whether they are the subject of the scenarios themselves or lie outside of their remit. In another context, the speculative infrastructures involved in oil exploration are now able to pinpoint fossil fuel reserves with a relatively high degree of accuracy. Highlighting the performative dimensions of this practice is not the same as questioning the reliability of the tools in question: It is simply to argue that the act of prospecting sets in motion a range of expectations, investments, and promissory statements that makes a future in which fossil fuels are extracted from the ground more likely.

If speculative infrastructures are coauthors, scriptwriters or stage-setters of the futures they envision, can the same be said of the temporal frameworks that underpin them? As some recent approaches to temporality within the humanities and social sciences have shown, the way that time itself is conceived has a bearing on the kinds of environmental futures that are imagined.[83] For although the idea of time as a singular, unfolding line from past to future is one upon which most people have come to rely, this does not mean that other temporal frameworks are not equally valid. Anthropologists such as Carol Greenhouse have argued that conceptions of time vary massively across and within cultures, and at moments of crisis there is often a power play in which one temporal framework seeks to establish legitimacy at the expense of others.[84] In Mark Rifkin's analysis, this hegemonic form of time is an "orientation" that has no more purchase on experiences of time than any other, constituting a form of "settler time" that has come to be imposed (often violently) onto other temporal orientations.[85] The Potawatomi scholar of environmental justice Kyle Powys Whyte has persuasively shown how shifting away from such settler temporal frameworks in discussions of climate change has the capacity to move political action away from ready-made solutions.[86] Instead of the frequently deployed idea of the climate crisis as a "ticking clock," which again has roots in Cold War doomsday imagery, Whyte advocates for an experience of time forged through interconnections, shared responses to environmental cues, and distributed responsibility—or what he calls "kinship time."[87] This kinship time does not lessen the urgency of climate change but rather reconceives urgency in a nonlinear way and, he argues, changes the kind of futures that might be imagined in response.[88]

Time, then, has a history. And as counterintuitive as it might seem, so too do futures. The year 2100 imagined in the 1960s is not the same year 2100 many imagine today. Several recent books explore this idea, taking a broad sweep of such cultural representations of futures and forming insights about the hopes and fears of the societies whose dreams they encapsulate. For example, Lawrence R. Samuel's *Future: A Recent History* charts the ebbs and flows in American attitudes toward futures, from the optimism of post–World War II reconstruction to the anxieties around futures swirling at the time of the Cold War.[89] Likewise, for Daniel Rosenberg and Susan Harding, the rich cultural history of futures goes to show that they are far from being "empty" or awaiting lived content. Instead, futures are perhaps even "overdetermined" as objects of experience, saturated with the imaginings of previous generations, such that today our futures have become "increasingly citational."[90]

Such histories of the futures inevitably weave their way through this book, but there is no attempt, as in some of the other books mentioned above, to pinpoint a spirit of the age or a general disposition toward futures belonging to a specific nation or decade. The focus is less on the insights that can be gleaned by aggregating cultural representations of *the* future and more on the tools, techniques and methods that act upon such futures, opening and closing corridors to create or restrict imaginative space.

PRESENTS—PASTS—FUTURES

One prominent argument concerning cultural attitudes toward futures that is worth engaging with (if critically) is the historian François Hartog's idea of "presentism," because it is explicitly contrasted to futurism.[91] In abbreviated form, Hartog's argument is that the relationship between the past, present, and future is not fixed but historically variable, comprising distinct "orders of time," or "regimes of historicity."[92] In each regime of historicity, the stress is laid in a different place. Whereas "the West has spent the last two hundred years dancing to the tune of the future," more recently, a 'sense of [a] permanent, elusive, and almost immobile present" has come to dominate.[93] The word "presentism," then, denotes an experience of time no longer oriented to the future but rather characterized by "ever greater speed and mobility or living from hand to mouth in an ever stagnating present."[94]

As can be inferred from some of the vocabulary Hartog uses to frame this discussion, the account is suffused with a nostalgia for a period in which looking ahead supposedly played a far greater role, until, that is, the "light beaming in from the future" began to wane and "what was ahead seemed increasingly unpredictable."[95] As historiographers such as

Chris Lorenz have pointed out, however, Hartog's views on presentism are themselves rooted in a particular historical context. Written against the backdrop of the so-called "memory wars" in France in the second half of the twentieth century, during which many historians lamented the rise of heritage, commemoration, and collective memory, Hartog's books and essays are also interventions into a debate aimed at preserving the alterity of the past and the openness of futures vis-à-vis the present.[96] For Hartog, a culture of commemoration emerged in France in the 1980s that placed the past in the service of shoring up group identity, a situation that, in turn, threatened the autonomy of history. This perhaps explains the Eurocentrism of the idea of presentism, a critique that could be extended to other thinkers such as Franco "Bifo" Berardi, who has also bemoaned the "slow cancellation of the future."[97] This lament is usually made by contrasting it with the kinds of progressive temporal experience that characterized Western modernity. The question then becomes what futures, and what modes or temporalities of cancelation, are at work (the mode, here, is quite distinct from those associated with infrastructural lock-in, path dependencies, and "defuturing" mentioned above). While the apparent dwindling of progressive energies in Western democracies has left many high and dry—particularly in former centers of industrial labor—it cannot be assumed that the same temporal experience is universally shared. Just as the concept of a monolithic modernity has unraveled in recent decades to account for the varied cultural experiences the term conceals, so too can the concept of a "regime of historicity" become more nuanced and multifaceted. As some of the anthropological accounts referenced above have shown, the temporal frameworks used to divide pasts, presents, and futures before they can be braided together again are even more variable than Hartog suggests, and are inescapably colored by ideological framings and the positionality of the theorists who speak of them.[98]

This temporal heterogeneity is further amplified by climate change, which, in the words of Andreas Malm, is characterized by a "messy mix-up of time scales."[99] To illustrate this, I could cite an example from my own immediate environment: the high concentrations of heavy metals currently found in peat bogs in parts of England, a country whose historical responsibility for climate change dwarfs its population size. Absorbed at the time by working-class bodies during the height of the Industrial

Revolution, the fumes from factories also drifted over moorland, where metals now leach into the water supply and form sediments that are so contaminated they need to be disposed of as toxic waste. The peat itself, which has been extracted from parts of the UK since neolithic times, is now being valued for its potential role in mitigating the effects of climate change. Projects such as "Moors of the Future" enlist volunteers to help restore peat bogs largely because of their capacity to act as a carbon sink for emissions yet to be created.[100] Here, the slow drip of toxic accumulation reaches a tipping point that triggers ecological restoration activities. The perceived necessity of this restoration is augmented by other perceptions concerning the future effects of climate change that can be offset in the present.[101] Meanwhile, by using peat-land restoration to offset pollution created elsewhere, the moors are brought into contact with yet other sites of temporal conjunction—a point of contact that seems more fragile by the day as regimes of carbon offsetting unravel. This example, and many others like it, makes it difficult to uphold the idea of presentism as anything like a totalizing force, or the all-conquering effect of late capitalism's death drive. As Kolinjivadi and coauthors put it, "the increasing non-linearity of climate tipping points and positive feedbacks stands testament to a temporality that does not align and cannot be made to tick to the Time of capital."[102] The term "environmental futures" should, then, be understood in a double sense: capturing both the range of human imaginings of futures shaped by climate change and biodiversity loss, and the myriad ways in which more-than-human forces possess future-making capacities of their own.[103]

One aspect of Hartog's argument that is easier to support is the identification of climate change as a multiplier of uncertainty about futures experienced in the present. The scramble to contain these uncertainties is one of the reasons for the rapid growth of financial instruments such as weather derivatives, catastrophe bonds, and other novel ways of transforming uncertainties into risks that can be bought and sold.[104] It is also part of the reason for the decline of a certain kind of predictive rationality discussed in chapter 1. In the face of spiraling uncertainties, efforts to transform futures into the objects of probabilistic risk management lend support to nostalgic claims that they are no longer the spaces of radical possibility they once were—with the word "possibility" signaling a generative power that "probability" structures and curtails by ascribing likelihood. And yet there is also a sense in which, precisely because of the uncertainties of the present, certain risk management tools marketed at companies and organizations fail to gain enough purchase on the futures they seek to contain. The points of slippage leave space for futures to re-emerge as realms of possibility, or, to repurpose a phrase from Barbara Adam, as a field of "patterned creativity."[105] Both in spite of and because of the overwhelming uncertainties that play out across unruly conjunctions of pasts–presents–futures, there is a sense of futures as both capacious and overdetermined, objects of management that perpetually extend the present and points around which hope and fears might crystallize into something new.

CHAPTER SUMMARIES, AND A NOTE ON METHODOLOGY

How is it possible to speak of heterogeneous attitudes to futures without flattening them into one overriding "order of time"? In the same way that a mass of cables or tracks might represent more than one infrastructure when disentangled, this book works on the basis that there are multiple futures, and numerous speculative infrastructures that complicate the idea of a single unifying image able to bring them together. Rather than a problem, this generative multiplicity should be embraced as something positive. In the environmental sciences, it has long been known that models average out the underlying temporal and spatial heterogeneity—or "patchiness"—of marine and terrestrial ecosystems.[106] This patchiness is, however, often what gives environments their dynamism, in contrast to monocultures such as fish farms or tree plantations. There is no doubt that this book also engages in its own forms of averaging and simplification, but its structural variability is nevertheless an attempt to embody the patchiness and patterned creativity of critical futures work.

Temporal and methodological heterogeneity should alert the reader to the fact that while *Futures in the Making* discusses many artworks and aspects of visual culture, it is not intended to operate as a work of art-historical scholarship. The narrow space-time coordinates of a book focused on one decade and geographical location would have limited purchase when confronted with the unruly temporalities of environmental futures. Nor does it claim to offer a complete overview of contemporary art that deals with environmental futures. Each chapter situates a speculative infrastructure in a particular context—for example, the development of scenario thinking in the context of Cold War military planning and the burgeoning peace and environmental movements, or forms of prospecting in the context of the extractive industries—but these contexts are often treated as points of departure that facilitate jumps in time and space. The aim is not to provide a comprehensive overview of a particular group of creative practitioners and the sociopolitical milieu in which they work but rather to explore ideological or epistemological affinities and follow through on the methodological implications of treating time and temporality themselves as something malleable and open to creative play. Bricolaging diverse examples and ideas in this way is a fundamentally pragmatic task too, insofar as it assembles resources for thinking environmental futures differently—experimenting while encouraging experimentation.[107] This approach unavoidably leaves many paths not taken, and consequently there are a great deal of relevant artworks that have been discussed neither directly nor by proxy. It nevertheless allows for moments of nonlinear connection—for example, in chapter 2, between the nineteenth-century US survey photographs of Timothy O'Sullivan and a recent artistic project on lithium extraction in the Democratic Republic of Congo—that tell us something about patterns and practices of relating to futures that endure over many years, and that are often both specific to the context in which they are deployed and global in their reach.

Artworks and visual artifacts often serve to isolate the characteristics of a particular speculative infrastructure and help elucidate how visual culture plays a part in frustrating or furthering its aims. The four chapters that comprise this book take such speculative infrastructures as their titles, addressing histories and futures of scenario planning, prospecting, engineering, and enacting. In different ways, the artworks that cluster around these modes of futuring contribute a vantage point that is intended to open new vistas of thinking about how futures are constructed. What follows periodically draws upon personal interviews with artists, intended to open such vistas beyond what public exhibitions, printed sources, and archive visits can offer. Armed with this material, many of the chapters begin by working through the implications of an artwork or set of artworks, and not by setting out a stall for a particular theory or method. If one of the core arguments of this book is that artworks can help widen corridors of the futures beyond what the dominant techniques of futuring allow, then forgoing a panoramic overview or systematic historicization in favor of artworks that open up material and discursive space is an attempt to embed this aim at the level of the book's methodology. The activity of exposing existing connections and forging new ones is necessitated by the sheer scale and complexity of an environmental "polycrisis" that is simply too big, too multifaceted, and too slippery to be amenable to synoptic cultural analysis.[108]

The first chapter establishes this approach, isolating the logic of scenario planning in a recent artwork and using it as an entry point to debates around the instrumentalization of the technique. Such instrumentalization could be aligned to Ahlqvist and Rhisiart's identification of a "utilitarian" tradition within futures studies, which has resulted in the contemporary dominance of foresight. Providing a backstory to this narrative, the chapter explores the roots of scenario planning in the 1950s and '60s, a period that witnessed a sustained attempt to transform futures work into a rational pursuit, particularly by the American "civilian defense intellectuals" of the Cold War, the aims of which were later opposed by networks of peace and environmental researchers that coalesced around the World Futures Studies Federation. Cybernetics, game theory, and systems analysis all played a part in helping to give prediction a veneer of scientific credibility, before scenario planning—an approach that was much more modest and intuitive—started to gain traction, eventually growing into one of the most ubiquitous methods for constructing futures to have emerged over the past sixty years.[109] Developed by one of the key architects of cold war military strategy, Herman Kahn, scenario planning maps a plurality of *possible* scenarios that inform decision-making, rather than attempting to predict one *probable* image of the future (for an influential diagramming of this relation, see chapter 1, figure 1.5). Futures were no longer thought to emerge from the present in a linear fashion but to unfold along a series of "branching points" that were seen to indicate moments where critical decisions were required.

Following a path from think tanks, futures studies networks, and corporations to climate science organizations such as the IPCC, chapter 1 tracks the explosion in popularity of scenario planning toward the latter half of the twentieth century, reaching such a point that Ziauddin Sardar, then editor of the journal *Futures*, could describe it as "devouring futures studies."[110] At every step of this journey, the chapter shows, the formalized knowledge infrastructure of scenario planning has been part of a broader cultural field of activity, with roots in Hollywood, theatre, and science fiction. These roots gave rise to trunks and branches throughout the twentieth century, with early environmental artists such as Helen and Newton Harrison actively incorporating scenario planning into their work. In its practical applications, scenario planning is often asked to simultaneously embody open-ended exploration of possible futures and generate solution-oriented responses, with some calling for greater use of "normative" or "prescriptive" scenarios.[111] Engaging with this contemporary context, the chapter considers artworks by the likes of Peter Watkins, Arata Isozaki, Lena Dobrowolska & Teo Ormond-Skeaping, and Juanita Schläpfer-Miller & Manuela Dahinden, that exemplify the exploratory aspects of this mode of speculation, treating scenario planning as an aesthetic practice, or "infrastructure of feeling." These examples are not considered aesthetic purely by virtue of their status as works of art, film, or architecture. They are positioned as aesthetic insofar as they transform futures into objects of embodied experience, rather than the products of an abstract technology of governance. Toward the end of the chapter, the frame is expanded even further, with a look at speculative cultural movements such as Afrofuturism, and in particular the work of the group Black Quantum Futurism, which both carries forward some of the social-justice motivations of early futures researchers and helps turn scenario planning's logic of multiplicity toward the past.

Many of the approaches discussed in chapter 1 have a formal identity in the canon of futures studies. By contrast, the subject of chapter 2 is more diffuse, and proposes the idea of "prospecting" as a mode of speculation that includes but is not limited to the geological analysis of a territory. In the last few decades, the narrow definition of prospecting has been expanded to include bioprospecting, in which genetic resources are transformed into proprietary knowledge, frequently at the expense of communities who have cultivated this knowledge over generations. Prospecting is therefore almost always extractive, but insofar as it involves a gamble on the profitability of a resource that has not yet materialized, it is also inherently speculative. In the context of environmental futures, it has been turbocharged by the green energy transition and fears over crop resilience. Taking recent discussions of the "extractive view" as its starting point, one of the principal aims of this chapter is to show how forms of visualization are bound up in this mode of speculation. It investigates how the search for resources generates a visual culture *of* prospecting and a visual culture *about* prospecting, whether through aerial views of resource frontiers, spectacular images that set expectations in motion, or "specimen views" that isolate objects of economic interest. Encouraging leaps in time and space, from the nineteenth-century survey photographs by Timothy O'Sullivan to contemporary work by the likes of Edith Morales and the group On-Trade-Off, the chapter demonstrates how artists repurpose and diversify the visual culture of prospecting, for example, by documenting the forces at play in the struggle over minerals deemed necessary to support electrification, or by investigating the methods by which genetic raw materials are turned into patentable commodities.

The third chapter zones in on practices associated with "geoengineering," a term that describes a number of vastly different approaches to tackling climate change, from stimulating plankton blooms in the ocean to increasing cloud reflectivity through stratospheric aerosol injections. What most of these proposals have in common is a shared conception of the planet as a giant bio-geophysical machine whose inputs, outputs, and feedback processes can be altered with the right technical know-how. For many still holding out hope for effective climate mitigation measures, geoengineering represents a hubristic enterprise and an excuse to carry on with business as usual. While remaining highly critical of techniques such as solar radiation management and carbon capture and storage, this chapter seeks to unearth the onto-epistemological foundations of geoengineering as a techno-capitalist project rather than to disaggregate and rank a list of the technologies that have been proposed. It explores the roles of cybernetic ecology, systems rationality, and the "macroscopic view" in the formation of geoengineering as an emerging knowledge infrastructure. It also looks at the ways in which artists have contributed to this mode of apprehending the planet. The master plans of Peter Fend and his Ocean Earth Development Corporation feature here, as does the more critical climate art of Tega Brain, Julian Oliver, and Bengt Sjölén, whose collective work pushes fantasies of AI-coordinated geoengineering initiatives to their logical, and often absurd, conclusions. Such work helps concretize the macroscopic view in ways that allow viewers to unpick its role in rendering futures as objects of experience in the present. Technologies of abstraction are further unpicked in a work of speculative landscape architecture for Mumbai proposed by Anuradha Mathur and Dilip da Cunha, which also pivots toward a more propositional mode. Without engaging with the subject of geoengineering directly, this project nonetheless points the way toward a mode of engagement with environmental futures that loosens engineering's reliance on predictive logics of control.

A very different approach to environmental futures plays out across the final chapter of the book. Here, a series of artistic projects is aligned to social movements that attempt to enact principles of degrowth, post-development, and *buen vivir*. What each of these movements has in common is a core commitment to stepping outside of an economic paradigm obsessed with gross domestic product (GDP), accompanied by modernist fantasies of relentless innovation. The visual culture of the growth paradigm is discussed in the context of a blockbuster exhibition at New York's Guggenheim Museum, curated by the architect Rem Koolhaus and his research and design studio AMO. *Countryside, The Future* is considered critically not only in terms of the economic models it presupposes but also with regard to the underlying temporal dynamics that reinforce them. In contrast to the exhibition's expectation-shaping focus on precision farming, indoor grow systems, and ever-increasing crop yields, the other projects discussed in this chapter do not have an immediately recognizable "futuristic" aesthetic. Artworks by the likes of Adrián Balseca and Kathrin Böhm that engage with solidarity economies and small-scale community projects might at first seem out of place in a book about environmental futures. Indeed, one of the arguments here is that such projects only appear as a return to the past when placed against the backdrop of an impoverished linear futurity of economic growth and the neocolonial logic of development. Such projects actively embrace temporal heterogeneity, exemplified in Balseca's work *La Abundancia* shown in figure I.3, for the

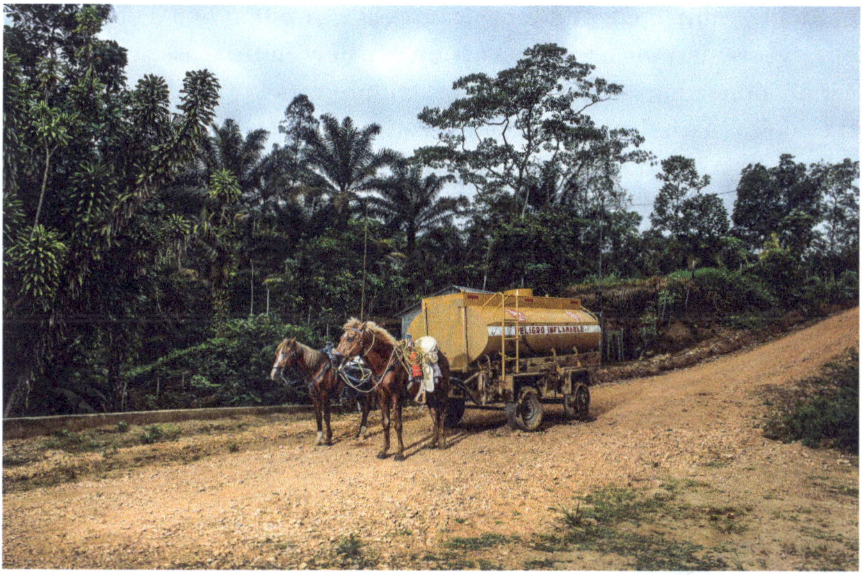

most part, through their mode of realization. Crucial here is the idea that post-growth futures can be "enacted" rather than simply imagined. Unlike some of the other modes of futuring discussed in this book, enacting is positioned as a form of speculation that actively embraces its performative power—in a manner not dissimilar from the "emancipatory" impulse within early futures studies—working toward the construction of spaces where the socio-ecological foundations of post-capitalist forms of life are cultivated.

The diffuse cultures of degrowth discussed in chapter 4, forged from unlikely alliances between activists, agroecologists, political economists, and artists, form the basis of a speculative infrastructure that is still in the process of emerging. Identifying it as an object of study has the capacity to reveal insights about how art and social movements come to be

FIGURE I.3

Adrián Balseca, *La Abundancia*, 2014, digital photograph. Courtesy of the artist.

mutually imbricated. And yet this process is not that of a dispassionate observer remarking on the formation of a new cultural assemblage. The act of connecting the relatively isolated projects in this chapter, and in the book as a whole, is inescapably motivated by a personal sense of their relevance. While this is not intended to be a book of autotheory, in which philosophical exploration is explicitly grounded in personal experience, it has nevertheless grown from a lifetime's exposure to a certain kind of anticipatory politics. Having lived through an era in which mainstream imaginaries of futures have either been enthralled by the techno-utopian visions of Silicon Valley or so fearful of socio-ecological crises that any deviation from hegemonic social orders is perceived as a risk, I began the research for *Futures in the Making* from a position of desire. This was, and still is, a desire to find ways to free my own imagination from this two-lane track, to find other routes out of the present, and to pass these reflections on to others. The act of gathering some of the artworks and cultural projects together into a book is intended to provide glimmers, not of hope, but of ways to enter new universes of reference that shape experiences of environmental futures in the present. To engage with futures as objects of study is, as Barbara Adam and Chris Groves note, to seek "an encounter with a non-tangible and invisible world that nevertheless has real and material consequences."[112] If one of the consequences of this book is to introduce readers to creative practices that spark curiosity, connection, and experimentation with different speculative methods, then it too will have contributed in a small way toward widening existing pathways to environmental futures.

1

SCENARIO PLANNING

To predict in social science is to lay the future in a straight-jacket.
—Johan Galtung, "On Future Research and Its Role in the World"[1]

INTRODUCTION

A small collection of unspectacular crops fills two greenhouses in the grounds of Zurich's old botanical gardens. Rather than devoting the growing space to rare cultivars, endangered species, or carnivorous oddities, the plants on display include a common variety of potato (*solanum tuberosum*), a variety of sweetcorn (*zea mays*), and two varieties of wheat (*triticum aestivum*, *triticum dicoccum*), all of which are known to grow in Switzerland with reliable success. Despite its unremarkable appearance, *Climate Garden 2085* (figure 1.1) is a space filled with drama, its contents charged with hope and fear for futures played out through the fate of individual plants as they adapt to a warmer environment. The interior spaces are fitted with temperature control systems, which maintain a constant +2 degrees Celsius above the average outdoor temperature in one greenhouse, and +4 degrees Celsius in the other. These two temperatures gesture toward different climate futures, the first informed by the United Nations Framework Convention on Climate Change's 2016 Paris agreement, in which 196 states committed to keeping global temperature rise to a ceiling of 2

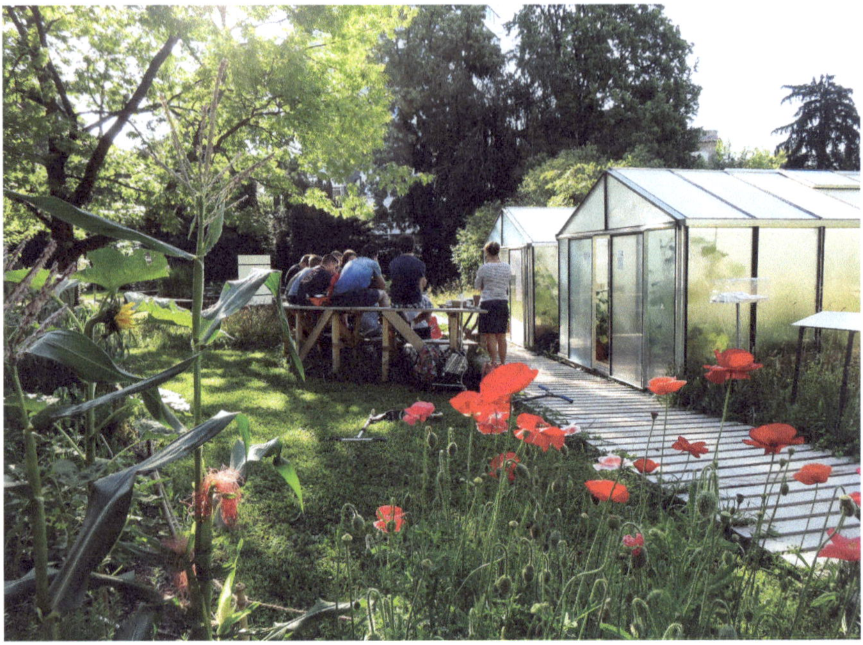

degrees Celsius above pre-industrial levels, and the second by a projection of the temperature in Switzerland in 2085 if warming continues at its present rate.[2] Further differentiations are made within the greenhouses themselves, with half of the plants subject to 30 percent less water, and the other half following a typical watering regime. A final collection of plants grow in the present, planted outside the greenhouses in soil that is carefully measured for temperature and moisture content.[3]

In making these small adjustments to temperature and watering, *Climate Garden 2085* reaches out toward a set of futures filled with drought, crop failures, and, in a minority of cases, increased yields. What is

FIGURE 1.1

Juanita Schläpfer-Miller and Manuela Dahinden, *Climate Garden 2085*, 2016. Installation at Zurich's Old Botanical Gardens. Courtesy of the artist.

significant here is that this public experiment trades not in the prediction of *one* future but in the mapping of multiple futures through scenarios that fan out from the present. The word "scenario" in this context has become so ubiquitous that it is easy to ignore. Now a common feature in climate change research and policy documents, and notably providing the underlying architecture for the IPCC's pathways, the role of scenario planning as a mediator of cultural relations with environmental futures has never been more important.[4]

To understand how scenarios operate today it is important to look backward as well as forward, tracing the genealogy of its forms to a Cold War milieu in which, as Joseph Masco notes, "a portrait of the planet as a fragile and integrated system" first emerged, albeit predominantly in the context of nuclear rather than climatological hazards.[5] Scenario planning grew out of the epistemic culture of systems theory, cybernetics, and operations research incubated in such organizations as the RAND Corporation; however, in contrast to the reigning predictive rationality of Cold War technoscience, promoted in different ways on both sides of the Iron Curtain, it announced a decisive break with the idea that futures were knowable as objects of rational planning.[6] Where other speculative tools developed around the same time, such as the Delphi method, aimed to produce one synthesized vision of the future, the emblematic form of scenario planning was one of branch points, decision trees, and a proliferation of alternate futures that were each, in themselves, internally plausible. Adopting an altogether humbler position, its proponents were keen to emphasize the subjective, even creative dimensions of the technique, describing it in terms associated with narrative and storytelling as much as scientific objectivity. "Scenarios," wrote Hermann Kahn in 1966, "are one way to force oneself and others to plunge into the unfamiliar and rapidly changing world of the present and the future."[7] They constitute an "explicitly structured set of assumptions, definitions, typologies, conjectures, analyses and questions."[8] A scenario, as Kahn and contemporaneous thinkers such as Daniel Bell repeatedly stated, is not a prediction but rather "an explication of possibilities."[9]

Since its formal development in the 1960s, the role of scenario planning has shifted back and forth along a spectrum that ranges from the pedagogic to the predictive, with *Climate Garden 2085* serving as an embodiment of this tension. Occupying a space somewhere between a science experiment and an artwork, its creators, Juanita Schläpfer-Miller and Manuela Dahinden, both of whom describe themselves as science communicators, frame the project as a "social learning experience" that runs counter to the "didactic futurology" of some science and public policy approaches to climate change communication.[10] As a now well-established interdisciplinary field, science communication contains within it a wide variety of approaches, ranging from those that privilege the unidirectional transfer of information from scientists to the public, to others that seek to develop new formats and languages capable of triggering social conversations and narratives around science.[11] In this work, scenarios are intended to facilitate conversations by making climate science amenable to aesthetic experience, translating the dry language of reports into "a narrative environment for holistic sensory immersion."[12] And yet, despite the intention to abandon didacticism, the scenarios used for *Climate Garden 2085* are also notable for their conservatism, following the IPCC's devotion to scenarios that are "most-likely" rather than using the tool as an opportunity to reflect on more radical responses to climate change.[13] By ascribing degrees of likelihood, the IPCC approach pulls scenario planning back toward prediction, eschewing the more exploratory and imaginative dimensions of scenario construction, which might facilitate socioeconomic pathways that don't rely on the assumption of continued capitalist innovation.

This chapter explores this tension at the heart of scenario planning by tacking back and forth between two aspects of its development. On the one hand, it discusses the systematic application of scenario planning by civilian defense intellectuals, corporate managers, and climate scientists, and on the other, it produces a sketch of what could be called the "expanded field" of scenario work, which includes creative practitioners working in film, architecture, and social practice. While only some in this second list are directly connected with one another, it is hoped that by bringing them together some affinities can be exposed. As we shall see, many of the creative practices discussed have the capacity to both bring the speculative infrastructure of scenario thinking to visibility and radically stretch its parameters. Readers expecting a compendium of environmental art projects in this chapter may be surprised by the diversity of concerns that enter this expanded field, from nuclear strategy to housing provision. While the underlying ambition is to shed light on how environmental futures come into being, at times this requires looking obliquely at cultural forms and processes produced with different goals in mind. The same could be said for the narrower, largely institutional Western history of scenario planning, which begins within think tanks such as RAND and the Hudson Institute, is taken up by corporate giants such as Royal Dutch Shell and the Boston Consulting Group, and eventually arrives as a truly globalized knowledge infrastructure in the present day. This is a history that has been told in some detail elsewhere (not least in the voluminous literature produced by practitioners of scenario planning themselves) as part of the broader field of futures studies, where it is supplemented by a fuller picture of how scenario planning was adopted by the peace and environmental movements.[14] What is only now being articulated is how this history is inextricably entangled with a wider creative imaginary of multiple futures played out in a global context and the implications this has on how environmental futures are shaped.[15] A central argument here is that scenario planning is not just a new descriptive mode of mapping multiple futures but rather an embodied aesthetic practice that can be felt as well as thought, or an "infrastructure of feeling."[16]

Toward the latter half of this chapter, these aesthetic dimensions come to the foreground, shattering any notion of the history of scenario thinking as one monolithic enterprise. In particular, the chapter concludes by considering the parallels and tensions between a history of scenario planning practiced in the worlds of corporate, scientific, and public policy research, and an expanded field of speculative thinking developed within cultural movements such as Afrofuturism. While the multiplication of futurisms (a non-exhaustive list might include Indigenous Futurism, Crip Futurism, Queer Futurism, Sinofuturism, Gulf and Arab Futurism) has become something of a trope in contemporary art and culture in recent years, unevenly alloyed with emancipatory political projects and social movements, an engagement with Afrofuturism introduces an important set of "temporal complications" that explicitly question the chronopolitical foundations upon which scenario planning relies, in ways that are more direct than in some of these other futurisms.[17] In other words, as well as documenting scenario planning's branching of possible futures from the present, the same logic is explored in reverse, exploring multiple pasts and recovering "histories of counter-futures" that take us beyond the typical methodological framing of scenario planning.[18]

"THE GENTLE ART OF REPERCEIVING"

Scenario planning was not formalized as a tool of futures studies until the 1950s, and yet the term "scenario" has a much longer history that stretches back to the sixteenth-century Italian commedia dell'arte.[19] Like the structures that underpin traditions in performing arts from many other cultures, here scenarios fulfilled the practical function of providing a basic outline of a play to the actors involved. Commedia dell'arte plays were principally works of oral improvisation that, unlike a scripted performance, might vary in execution from night to night. The scenario provided the basic scaffolding upon which this improvisation could hang, and included stage directions and descriptions of what was to happen in the play, and occasionally noted the comedic elements.[20] Adopting many of the conventions of theatre, the early cinematic experiments of the late nineteenth century also employed scenarios, albeit adapted for a format that had less room for multiplicity (in the obvious sense that a film cannot be changed with each screening, and insofar as the cost of film limited multiple takes and extensive editing).[21] While the scenario in this context was primarily a technical document, rather than anything deemed of artistic merit, it nevertheless provided cues, parameters, and temporal markers that permitted multiple interpretations.[22] With the creation of the studio system in Hollywood and the efficiency savings it introduced to film production, the scenario became increasingly important, and by the 1950s, it had slowly morphed, first into the "continuity script" and subsequently into the "screenplay."[23]

With the RAND Corporation based in Santa Monica, its preeminent military strategist Herman Kahn spent much of his time working in the shadow of the Hollywood Hills, and this physical proximity seems to have facilitated the transfer of a kind of "scenario thinking" from the entertainment industry to what can be considered the archetypal think tank. Chermack and Lynham have suggested that Kahn adopted the name for his new futurological method around the time it was superseded in Hollywood by the screenplay.[24] Another account places the humorist, screenwriter, and political scientist Leo Rosten as the bridge between the two worlds of the film industry and the Cold War think tank, and the person who directly suggested the term to Kahn, who apparently adopted it "because he liked the emphasis it gave, not so much on forecasting, but on creating a story or myth."[25]

It was this storytelling function that Kahn would apply to problems of future military engagements in his policy work, also primed by a childhood interest in science fiction (SF) literature. According to his biographer, Kahn remained an avid reader of SF throughout his life, and it would be surprising if he had not come into contact with the post-apocalyptic imaginaries of Isaac Asimov (in particular his 1950 novel *Pebble in the Sky*, in which earth is depicted as a planet whose inhabitants adapt to increased levels of radiation) or Nevil Shute's 1957 book *On the Beach*, first adapted for film in 1959, which also peddles a future characterized by nuclear apocalypse.[26] Despite his grounding in SF, Kahn was nevertheless clear that the kind of unrestricted speculation found in the genre should be subjected to a clear principle of relevance in scenario planning. Ghamari-Tabrizi's interview with Kahn's collaborator Anthony J. Wiener is instructive in this respect: "Herman would say, 'Don't imagine that it's an arbitrary choice as though you were writing science fiction, where every interesting idea is worth exploring.' He would have insisted on that. The scenario must focus attention on a possibility that would be important if it occurred."[27]

Wiener's casual observation here brings a key feature of scenario planning into view: that the likelihood of something occurring was not necessarily what made it a good candidate for a scenario; what was more important was the impact this event could have if it did occur. Plausibility was not insignificant to Kahn, but as he and Wiener wrote in their co-authored book, "It is important not to limit oneself to the most plausible, conventional, or probable situations and behavior. History is likely to write scenarios that most observers would find implausible not only prospectively but sometimes, even, in retrospect."[28] In this way, scenario planning represented a rejection of realism, epitomized in Kahn's famous imperative for Cold War decision makers to "think the unthinkable" and actively imagine what life would be like in a world traumatized by nuclear conflict.[29] Analysts at RAND irreverently disregarded the battle experiences prized by military generals, and Hermann Kahn was vocally dismissive of "on the ground experience," claiming that "the unrealized and unexperienced, but historically plausible, problems of World Wars III and IV are more valuable than the experienced problems of World Wars I and II."[30]

The book that journeyed furthest into the unthinkable was *On Thermonuclear War* (1960), which plunged readers into a range of alternative futures characterized by cataclysmic events that included evacuations, mass starvation, and multigenerational genetic defects arising from nuclear fallout. Scenario planning was the method that provided the underlying architecture for the book, although as befits any knowledge infrastructure, its role was largely overshadowed by the content it subtends—in this case terrifying visions cast in the cold light of technocratic indifference.

Ghamari-Tabrizi describes *On Thermonuclear War* in terms usually reserved for literary art, framing Kahn's style as "an experiment in historical narrative. His zigzags, pockets of scenario detail, and lists resembled the herky-jerky pulse of an epic poem."[31] Nevertheless, it would be the self-assigned task of a film director to imbue Kahn's future visions of nuclear apocalypse with the terror they deserved. Peter Watkins's *The War Game* (1965) was an unflinching look at the consequences of nuclear conflict on the south of England, and although commissioned by the BBC, the corporation deemed it too inflammatory to show on national television—a decision that violated their own charter of political neutrality insofar as it involved prior consultation with the British government.[32]

The War Game is singular in unspooling the consequences of a nuclear strike on the population of the UK, and its exploration of a near future is structured around exploring one scenario in all its gory detail. Exploring the future in the singular, I would argue, does not automatically imply the film operates in the register of prediction. Much of the action and commentary is voiced in the future conditional tense, a grammatical form proper to scenario planning.[33] At one point, a voiceover in the style of a newsreader makes this explicit, overlaying a scene in which a man argues with a police officer about having to take in eight evacuees with the statement that "should Britain ever thus attempt the evacuation of 20 percent of her entire population, such scenes as these would be almost inevitable."[34] Other sections of the film depict widespread death and disease, the armed looting of government food supplies, and execution by firing squad as the police seek to impose order—in short, a series of events entirely in keeping with the kinds of post-strike scenario Kahn sketched out in his work at RAND.

By then a global celebrity, it is perhaps unsurprising that references to Kahn pepper Watkins's film. A range of talking heads from the present—some, members of the public; others, establishment figures such as priests and politicians—are intercut with scenes of a future nuclear holocaust, causing the film to jump between two different time periods. An intertitle announces that one such monologue is "based on the recorded statements of a leading American Nuclear Strategist," and a figure closely resembling Kahn appears on screen to make the following claim, to almost comic effect: "In the next World War, I believe that both sides could stop before the ultimate destruction of cities, so that both sides could retire for a period of ten years or so of post-attack recuperation, in which World Wars IV through VIII could be prepared" (figure 1.2).[35]

Although Watkins's film would not reach its intended audience on British television until 1985, its commissioning twenty years before is indicative of the influence of scenario thinking in the 1960s. Working in quite a different register, although one perhaps no less cinematic, the celebrated Japanese architect Arata Isozaki also created images of future nuclear destruction. In *Re-ruined Hiroshima* (1968, figure 1.3), Isozaki collaged images of giant skeletal megastructures on top of an elevated view of the decimated cityscape, photographed shortly after it was bombed in 1945. While the additions belong to an architectural language of the future, linked to the Metabolist movement with which Isozaki was associated, these megastructures also appear in states of disrepair, inviting reflection on the destiny of all architecture to one day fall into ruin. Installed as part of the exhibition *Electric Labyrinth* at the 14th Milan Triennial, the piece was incorporated into an elaborate exhibition design in which hanging panels slowly rotated when a visitor passed through an infrared beam. A slide carousel projected hundreds of other future buildings onto the collage, representing the construction and ruination of countless architectural visions.[36] This was a complex and layered vision of futures, in which

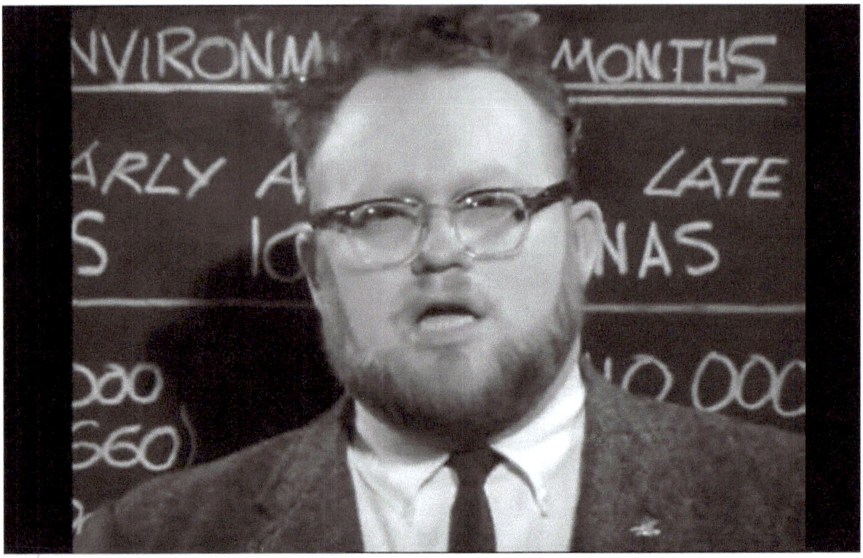

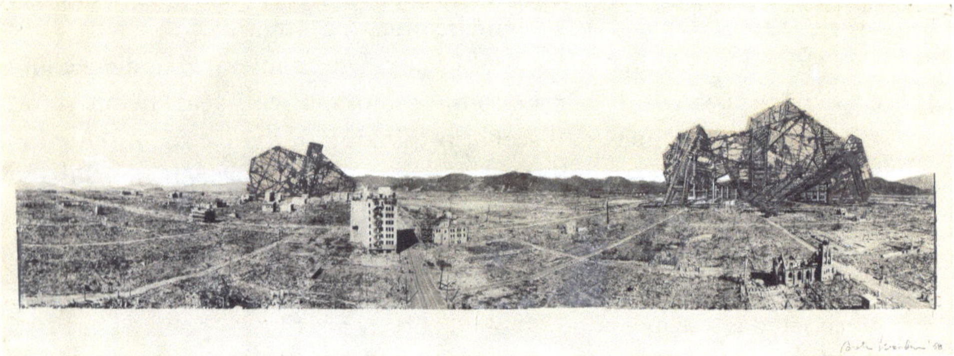

FIGURE 1.2

Still from *The War Game*, featuring a character with striking resemblance to Herman Kahn. Directed by Peter Watkins (London: BBC, 1965), 21:23.

FIGURE 1.3

Arata Isozaki, *Re-ruined Hiroshima*, 1968. Digital image, The Museum of Modern Art, New York/Scala, Florence.

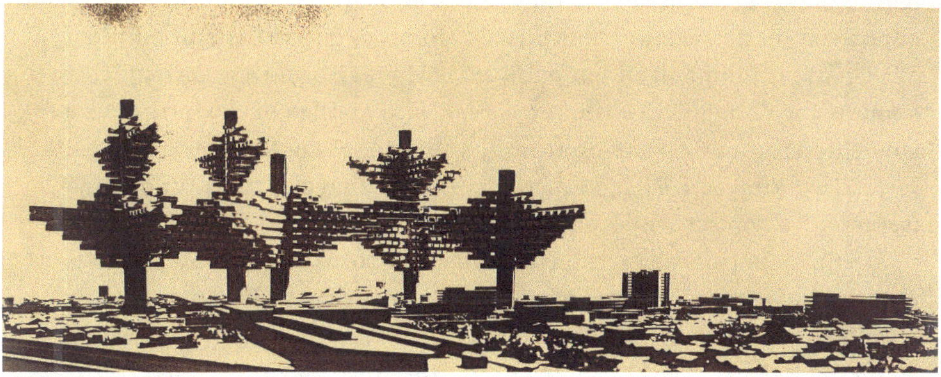

spatial and temporal scales were rendered malleable, but crucially it was also one that operated in dynamic interaction with the audience, rather than as a stable, pre-packaged vision of a future yet to come. The audience interaction was perhaps more dynamic than some had expected, with the student mobilizations of 1968 indiscriminately destroying some parts of the Triennial on the opening night. When a letter calling for support of the student movement was circulated in the following months, to the surprise of the exhibition organizers it bore Isozaki's signature—because, he later reflected, he thought this work "had precisely the same mentality, the same direction, as the student movement at the time."[37]

As well as the experiential aspects of Isozaki's nuclear ruinscapes, it is worth reflecting on how the installation was produced only a few years after other techno-utopian visions by the architect, such as *Cities in the Air* (1962, figure 1.4). This design also resisted stasis, featuring buildings in modular tree-like structures that could be modified or expanded—embodying one of the central tenants of Metabolism as an architecture responsive to human vitality.[38] Reading the two projects dialectically and

FIGURE 1.4

Silkscreen print of Arata Isozaki, *Cities in the Air*, 1991/1962. © Estate of Arata Isozaki.

in the context of a broader cultural shift away from the linear imaginaries of postwar modernization, it is possible to discern here the embodiment of scenario thinking in all but name. Elsewhere, Isozaki speaks explicitly about using "imaginative time" to break with an idea of the future "as a single determinate extension from the present along this uniform axis," instead working with a lived experience of time that is "arbitrary and indetermined," a point to which we will return toward the end of this chapter.[39]

In their speculative architectural visions, Isozaki and the Metabolists were part of a broader surge of interest in futures in 1960s Japan. By the middle of the decade, preparations were already underway for what would go on to be the first world's fair in Asia: the 1970 Osaka Expo, led by the architect Kenzo Tange, who invited many young architects to take part. Building on his work at the Milan Triennial, Isozaki would design the event system for the festival's central plaza, featuring two 14-meter-tall robots that could emit light, sound, smoke and even smells. Flexibility was again the guiding principle, and many of the lighting, stage, and seating components could be rearranged to accommodate different events. The theme of the expo was the future-orientated notion of "Progress and Harmony for Mankind," and the effect of Isozaki's contribution was to make the audience feel these futures in his cybernetically controlled sensory environment.[40]

1970 would also be the year that the newly formed Japan Association for Future Studies (Nihon Mirai Gakkai) hosted the second International Future Research Conference, spearheaded by the Norwegian peace theorist Johan Galtung. Although a much smaller event than the vast world's fair, the conference in Kyoto is significant in the history of futures studies as a milestone that contributed to the creation of the World Futures Studies Federation (WFSF), an organization whose social justice motivations were diametrically opposed to much of the futures work carried out in think tanks such as RAND.[41] The link between these events is not only one of synchronicity, however. As architectural and cultural historians William O. Gardner and Hyunjung Cho have documented, the two networks overlapped in some significant ways. The thematic kernel of the Osaka world's fair was shaped by a study group under the name "Thinking the Expo," and two key members of the Metabolist group, Kurokawa Kishō and Kikutake Kiyonori, were both in attendance at the conference in Kyoto.[42] Conceptually, both the Metabolists and many futures researchers were thinking through issues of both the atomic age and environmental catastrophe, with the decision to host the 1970 conference in Japan providing a symbolic backdrop twenty-five years after the attacks on Hiroshima and Nagasaki.[43] In his address to the conference, Galtung explicitly likened what he called "prescriptive futures work" to architecture, insofar as it "stipulates values to be realized, then goes ahead to do the job, on the drawing table (the theory, the blueprint), at the level of the model (the utopian community), in full scale reality (as an effort at large scale implementation)."[44] Whatever the scale, futures work required thinking in the multiple. Only then could the illusion of an objective and unbreakable line between past–present–future be overcome.

As both Watkins's film and Isozaki's architectural environments attest, the dynamic between Cold War military planning and the broader cultural sphere of art, film, and literature is not simply one of diffusion, according to which ideas pioneered by a Cold War avant-garde of civilian defense intellectuals slowly percolate through wider society. Nor are the exchanges between scenario planning as it exists within Cold War research and this expanded cultural field limited to the circulation of the word "scenario" itself. What was formalized as a technique by Kahn and subsequently consolidated into a knowledge infrastructure by countless others had deep roots in theatre, Hollywood, and science fiction, and would exist in reciprocal relation to art and visual culture for decades to come.[45] While the tools at the disposal of Cold War strategists were becoming less deterministic—in the West through scenario planning, and in the USSR through alternatives to forecasting and central planning pioneered by the likes of Georgy Petrovich Shchedrovitsky—their dispassionate framing largely reduced the experiential outputs of scenario planning to the language of the policy document and internal report.[46] By contrast, figures such as Isozaki and Watkins, whether directly influenced by the formal development of scenario thinking or otherwise, provided opportunities for audiences to feel possible futures, rather than simply "think the unthinkable." In parallel development to the knowledge infrastructures developed within organizations such as RAND, outlines of an "infrastructure of feeling" in artworks and films that deal with planetary futures are discernible. As the sociologist Daniel Bell wrote in 1964, society as a whole had become "future orientated," and yet the cultures of looking ahead that accompanied this orientation played out very differently across the political spectrum.[47] The following section looks at the diverse individuals and organizations that took up scenario planning in the 1960s and '70s, before the discussion turns back to environmental futures and creative practice in the second half of the chapter.

SCENARIOS: BETWEEN THE CORPORATE AND THE CRITICAL

Well before the publication of *On Thermonuclear War*, scenario planning was already established enough within the domain of military planning for United States Secretary of Defense Robert McNamara to begin using the procedure soon after he took office in 1960.[48] The following years mark a period in which scenario thinking expanded greatly from its position within the Cold War state apparatus, finding areas of application within social and corporate planning, as well as the burgeoning peace and environmental movements, as they converged in organizations such as the WFSF. An important vehicle for this expansion was the Commission on the Year 2000 (CY2000), which, under the direction of sociologist Daniel Bell, brought scenario planning, together with the more prediction-orientated Delphi method, to a much wider realm of application.[49] It is from this institutional backdrop that two of the most enthusiastic applications of scenario planning in the 1960s emerged: Kahn and Wiener's *The Year 2000* and an edited book with the very similar title *Toward the Year 2000*, which chronicled some of the organization's approaches.[50] As a means to introduce scenarios to the corporate world, Kahn established a three-part program at his new think tank, the Hudson Institute, that initiated company executives in the practices of planning for multiple futures.[51] The list of organizations that took up this opportunity, as the cultural theorist R. John Williams has shown, included over sixty global corporations working in a wide range of sectors, from information technology (IBM, Xerox) and automobiles (Ford, Volvo, Nissan), to banking and insurance (National Bank of Mexico, Prudential).[52] Later, other consulting firms, such as the Global Business Network and Boston Consulting Group, would spread the practice even further.[53] Crucially, it was also popular among oil companies, in particular Royal Dutch Shell, who would use the tool to dramatic effect in the 1970s.

Melinda Cooper explains this upsurge of interest in scenario planning using the concept of "turbulence," which is understood to operate both in meteorological terms and metaphorically as a way of characterizing uncertainty.[54] Cooper suggests that as the postwar international economic order established at the Bretton Woods conference in 1944 came to an end, so too did a period of relatively stable trade between nations, conducted using currencies backed by the gold standard. Around the same time that a single monetary reference point was being replaced by multiple, shifting determiners of value, the power structures that influenced the price of oil were also multiplied with the formation of OPEC in the early 1960s. As a result of the shift to fiat currencies and an increasingly unpredictable global trading environment for oil and many other commodities, there was an urgent need for tools that would facilitate decision-making under conditions of uncertainty. These tools, according to Cooper, were built upon "a model of strategic power that aspires to harness 'whatever' happens, . . . a power of all possible worlds."[55]

Even before their embrace of scenario planning, there was a sense in which Royal Dutch Shell was already attuned to the significance of this global shift toward managing uncertainty. Within its London headquarters, it established a division dedicated to "Long-Term Studies," in which Ted Newland recalls he was "told to think about the future, with no real indications of what was required."[56] Visiting Kahn in 1966, a year after the division's founding, Newland found himself one of two representatives of the company who attended such meetings; the other, his French counterpart Pierre Wack, would soon be transferred to the company's London headquarters.[57] Bringing their new speculative tools back with them to Shell, the two men led a team that would prepare detailed scenarios dealing with a range of possible shifts in world commodity markets. In an account that would be retold in management literature several decades hence, Newland and Wack's scenarios of 1971 and 1972 suggested one possible future in which the power of oil-producing nations to determine prices eclipsed that of consumers, preparing managing directors for an eventuality that would come to pass merely a year later with the 1973 oil embargo.[58] Over the following two years, Shell would climb the corporate rankings, from the eighth- to the second-largest oil company in the world.[59]

Whilst many directly attributed Shell's ascendancy in the 1970s to their uptake of scenario planning, the reality is more complex. The company's experiments with the new method did not predict the oil crises of the 1970s, scenario planning's proponents maintained, for business forecasting and scenario work were two fundamentally different activities. As a best-selling business studies book on scenarios asserts, using the approach simply "allowed the company to override the domination of the credible, popular but very wrong imagined future."[60] As Wack put it, instead of trying to produce increasingly sophisticated forecasts, the correct approach was "to accept uncertainty, try to understand it, and make it part of our reasoning."[61] Shell maintains this commitment to the openness of scenario planning to the present day. In a recent report dealing with Germany's future energy system, the process is described in pedagogic terms as "designed to stretch management thinking and even to consider events that may only be remotely possible."[62] Scenarios do not, the report suggests, provide solid foundations upon which investment decisions can be made; rather, they expand the perceptions of those responsible for making decisions, and order these perceptions into a range of bounded possibilities. This subtle but important distinction decisively shifts the focus away from producing one *probable* image of the future to mapping an array of *possible* scenarios that inform decision-making, as shown in the so-called "futures cone" diagram produced by futurist Clem Bezold and the health researcher Trevor Hancock (figure 1.5).[63] What this diagram doesn't capture is the complexity with which the straight lines between the present and the future(s) can bifurcate, break, and loop back on themselves within this speculative infrastructure, a set of further complications that will be explored in more detail in the last section of this chapter.

And yet, despite the guarded language of Shell's reports, there is a sense in which the investment decisions of the 1970s were inevitably guided by the scenarios they contained. As confidence in prediction wavered, scenario planning seemed uniquely positioned to answer the demands of corporations for a tool that would enable them to keep profits accruing even in circumstances wherein long-range forecasting was

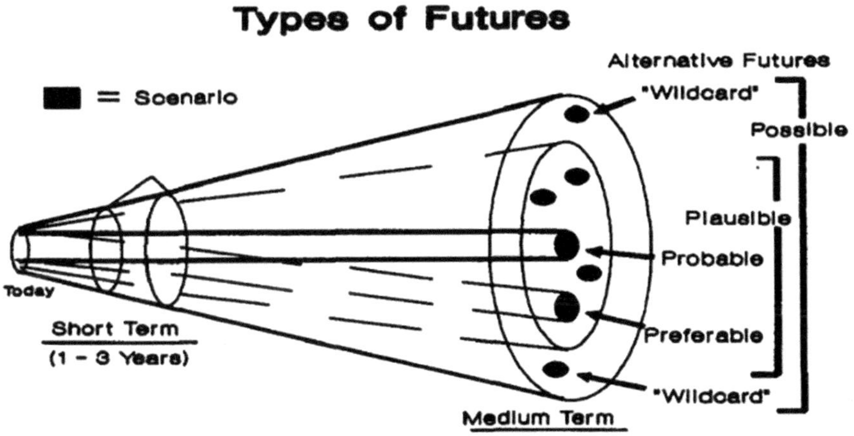

deemed impossible. For Wack and others at Shell, uncertainty was "not just an occasional, temporary deviation from a reasonable predictability; it is a basic structural feature of the business environment," and one that requires a new supply of tools and methods. In this respect, the rapid growth of financial instruments such as derivatives during the same period is noteworthy, as they enable the hedging and spreading of bets across multiple futures. Like the scenario, the grammatical tense of the derivative is the future conditional, and together they represent mutually reinforcing responses to the new atmosphere of uncertainty in the late '60s and '70s, which would only be exacerbated as climate change entered public consciousness in the decades that followed.

FIGURE 1.5

Clement Bezold and Trevor Hancock, Futures Cone Diagram, 1993. From Clement Bezold and Trevor Hancock, "Introduction and Overview of Health Futures," appendix 5 in Ann Taket, *Health Futures in Support of Health for All: Report of an International Consultation, Geneva, 19–23 July 1993* (World Health Organization, 1993).

Cooper's arguments concerning turbulence run against the grain of those keen to position scenario thinking as an intrinsically open, if not utopian mode of relation to futures. Management strategist and former CEO Clem Sunter is one of many people who make grand claims for scenario planning, arguing that it helped end apartheid in South Africa, because "it changed the conversation of an entire nation."[64] Its development within centers of state and corporate power also provides an important qualifier to the arguments of cultural theorists such as T. J. Demos, who are keen to emphasize the political radicalism of artworks and cultural projects that "escape homogenous time" by producing "a future of many futures."[65] While Demos is attentive to the "chronopolitical diplomacy" necessary to mediate between multiple futures and the diverse temporal models upon which they rely, the history of scenario planning shows that, when disciplined into a method, there is nothing intrinsically radical or emancipatory about the capacity to think futures in the multiple; it can just as easily serve as a tool to ensure capitalist hegemony.[66]

In the 1960s and '70s, scenario planning nevertheless crossed ideological divides. At the same time it was being embraced by corporate giants such as Shell, many within the nascent WFSF also showed enthusiasm for the technique. In the proceedings of the inaugural International Future Research conference in 1967, for example, convened before the WFSF was formed, the peace campaigner Robert Jungk posed the question of whether scenario writing, along with a host of other approaches, could be put to progressive ends.[67] At the same event, Galtung heaped praise on work by Kahn and Wiener that used the method, and, a few years later at his Kyoto address, argued that futures research should be orientated around a "genuine pluralism," since "any other direction only leads to an impoverished future which can easily be petrified."[68] Central to the embrace of scenario planning by activist-intellectuals such as Jungk and Galtung was its role in supplying "images" that could trigger the imagination to think outside the constraints of the dominant system. Here, "images" of futures are not necessarily direct representations through art or visual culture (though these can play an important role) but rather diverse crystallizations of ideas and values that can help guide society toward desired outcomes. They embody what, in the introduction, I called the "generative multiplicity" of futures research, beyond an overriding image of success in a competitive environment that underpins corporate applications of scenario planning. In making this argument, Jungk and Galtung connected their ideas to that of Fred Polak, whose influential work was translated by the sociologist, peace activist, and Quaker Elise Boulding; Boulding and her husband, Kenneth, later incorporated Polak's work into theirs.[69] As Jenny Andersson argues, this constellation of futures researchers in the 1960s and '70s, provided a kind of "counter expertise" that repurposed many of the tools incubated within organizations such as RAND to explicitly normative ends.[70] Stripping away claims to scientific objectivity, many of the futures researchers that gathered around the WFSF were enthusiastic in putting futures work in the services of peace, equality, and environmental activism.[71] While the full story of the WFSF and the many other organizations that engaged in futures work is outside the remit of this chapter, it's worth noting that, in subsequent decades, the apparent gulf that separated the "emancipatory tradition" of futures studies from centers of state and corporate power was steadily closed, not least of all in what would later be known as the "Californian ideology" that blended countercultural politics, spiritual enlightenment, and entrepreneurial ambition.[72]

Situating the growth of scenario planning in this wider historical and socioeconomic context allows us to see that it is not wedded to any one ideological position. At the same time that it helped concretize images of futures for the peace and environmental movements, it also enabled corporations such as Shell to modulate sequences of events to steer them in a direction amenable to ever-growing profits.[73] In line with the rise of corporate foresight and its competitive applications, however much scenario planning was able to stretch the imaginative capacity of Shell's upper management, it did not prompt them to envisage the possibility of acting in a way that would place the wellbeing of the environment and people before profit.[74] As early as 1989, the company produced a "Sustainable World" scenario, in which global society shifts toward an outlook considered "seriously green."[75] Despite the fact that all top-level decisions had to be justified against the company's global scenarios, including this one, the "seriously green" vision did not end its toxic plundering of areas such as the Niger Delta,[76] which was awash with accusations against Shell of complicity in the murder of those fighting to protect the environment, such as Ken Saro-Wiwa and eight others executed by the Nigerian state.[77] Nor has the company's internal acknowledgement of their contribution to climate change as early as 1991 prevented the continued expansion of its fossil fuel infrastructure, including billion-dollar investments in Arctic oil drilling.[78]

THE BIRTH OF CLIMATE SCENARIOS

At roughly the same time Shell was riding out the 1973 oil crisis, two artists in California were making one of the first artworks to deal directly with climate change. Having already committed themselves to ecological issues, Helen and Newton Harrison had stumbled upon two publications that seemed to point toward very different futures. The first, by Reid Bryson, argued that the world was currently enjoying an interglacial period, and that it would not be long before a great chill descended on the planet once again. The second, by the Canadian physicist Gilbert Plass, was one of the first to correlate CO_2 emissions with global warming, foretelling a future in which the planet would warm by 1 degree Celsius by the year 2000 if emissions continued to rise at the present rate.[79] Summarizing this research in handwritten notes, the work overlays text that is part proposition, part poetry, on top of a cartographic projection of the world with San Diego at its center (figure 1.6).[80]

The sources the Harrisons used for the work were drawn from different scientific fields, operated across different time-scales, and proposed radically divergent futures. One thing they had in common was that they were predictive, and in Plass's case, predictive with remarkable accuracy.[81] Through a simple act of summary and combination, the Harrisons transformed these predictions into scenarios, opening up an array of futures that can be apprehended (if not inhabited) concurrently. This is reflected in the title of the piece, *San Diego as the Center of a World* (1973), in which the artists' hometown serves as a linguistic and visual anchor connecting multiple future worlds to the here and now (the emphasis very much being on *a* world, rather than *the* world). The branching of possible futures is also positioned as the object of planning in the present, and the last of the five paper sheets that comprise the work contains a number of "if . . .

then . . ." statements, culminating in a proposition that, in the absence of certainty around which future would come to pass, both long and short-range planning could begin.

The emphasis on chance and indeterminacy in *San Diego as the Center of a World* sets it apart from other environmentally inflected work of the time—in particular the numerous environmental art projects from various parts of the world informed by systems theory, which often worked on the basis of closed loops and self-regulating systems.[82] As the duo reported, the piece began from the chance discovery of Bryson's and Plass's publications during a "wander in the library" fueled by Helen's "noetic powers," rather than from a purposeful research agenda.[83] Although not unprecedented among artworks of the time—artists associated with Fluxus had employed chance procedures throughout the '50s and '60s—there is a sense in which the Harrisons were using chance and indeterminacy not to reduce their executive agency in the creation of artworks but rather to register the multiplication of future uncertainties already suggested in the emerging literature on climate change available at the time.

One publication that captured this sense of uncertainty in a scientific context was Hermann Flohn's paper "Climate and Energy: A Scenario to a 21st Century Problem," published in 1977.[84] This can be considered the first scientific article to use scenario planning to think about climate change. Flohn presents three scenarios, two of which closely resemble those upon which the Harrisons drew in their artwork, and makes an implicit reference back to the dramaturgical origins of scenario planning by structuring his reflections into three acts. Sketching these diverse climate futures, Flohn's ambition is to provide a survey of the field, "to ventilate . . . some possibilities of near-future climatic evolution based on the most recent state of knowledge."[85] As the geographer Mike Hulme has pointed out, Flohn's subjectivity as an author is deliberately foregrounded in places, in stark contrast to the language of the many climate scenarios produced since.[86] For Flohn, the sketch he presents is not objective, but rather "a perspective, a scenario outlined after many discussions among many leading experts."[87] In both cases, scenario planning is an activity with clear aesthetic dimensions, insofar as aesthetics is bound up with problems of perception and feeling at the root of the word.[88]

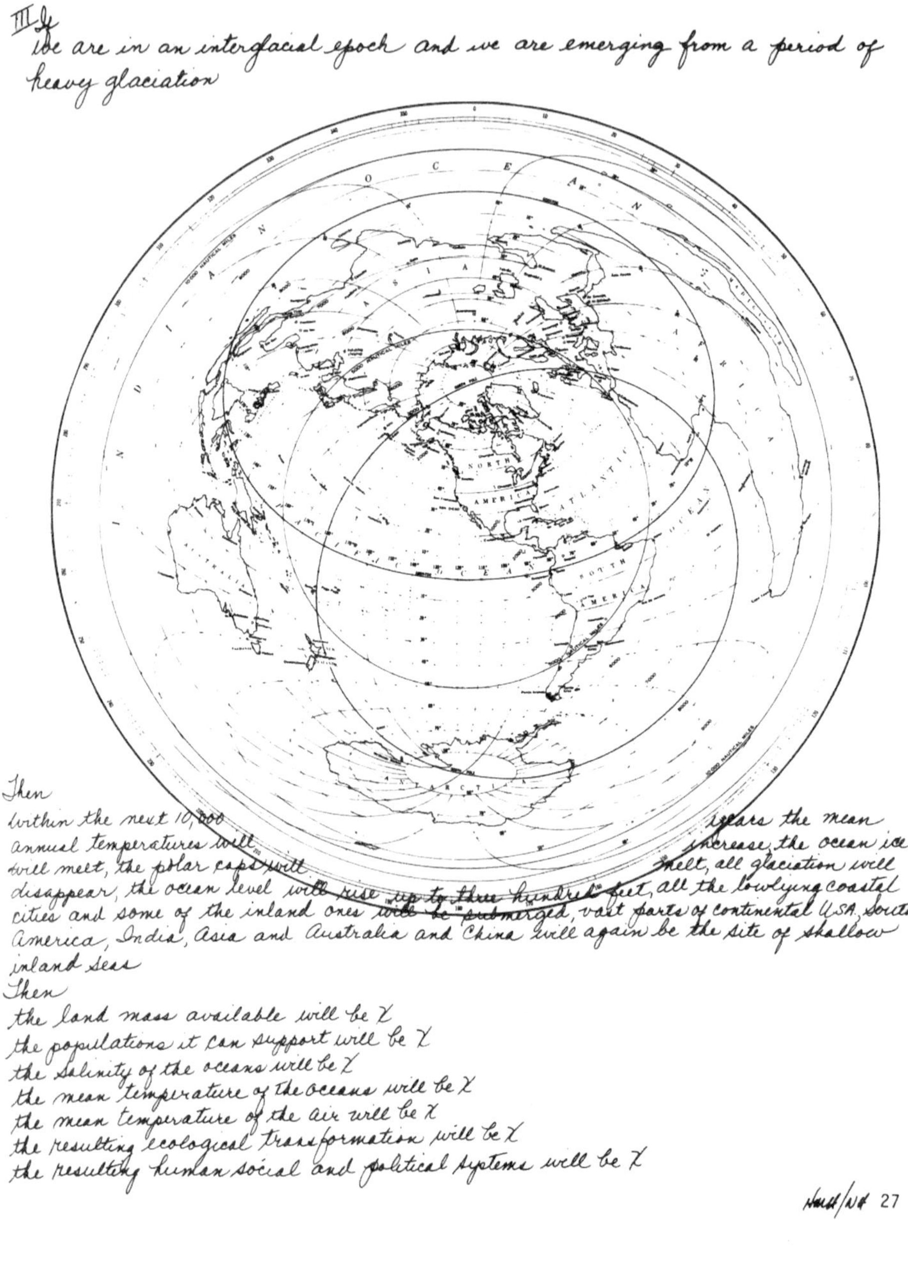

III If
we are in an interglacial epoch and we are emerging from a period of heavy glaciation

Then
within the next 10,000 years the mean annual temperatures will increase, the ocean ice will meet, the polar caps will melt, all glaciation will disappear, the ocean level will rise up to three hundred feet, all the lowlying coastal cities and some of the inland ones will be submerged, vast parts of continental USA, South America, India, Asia and Australia and China will again be the site of shallow inland seas

Then
the land mass available will be X
the populations it can support will be X
the salinity of the oceans will be X
the mean temperature of the oceans will be X
the mean temperature of the air will be X
the resulting ecological transformation will be X
the resulting human social and political systems will be X

27

FIGURE 1.6

Helen and Newton Harrison, *San Diego as the Center of a World*, 1973, pages 3–4 of 5. Courtesy of the Helen and Newton Harrison Family Trust.

IV. I.
According to Plass, the burning of the currently known supplies of coal and oil over the next hundred years multiplies the CO_2 tonnage of the air 18 x, increasing the green-house effect

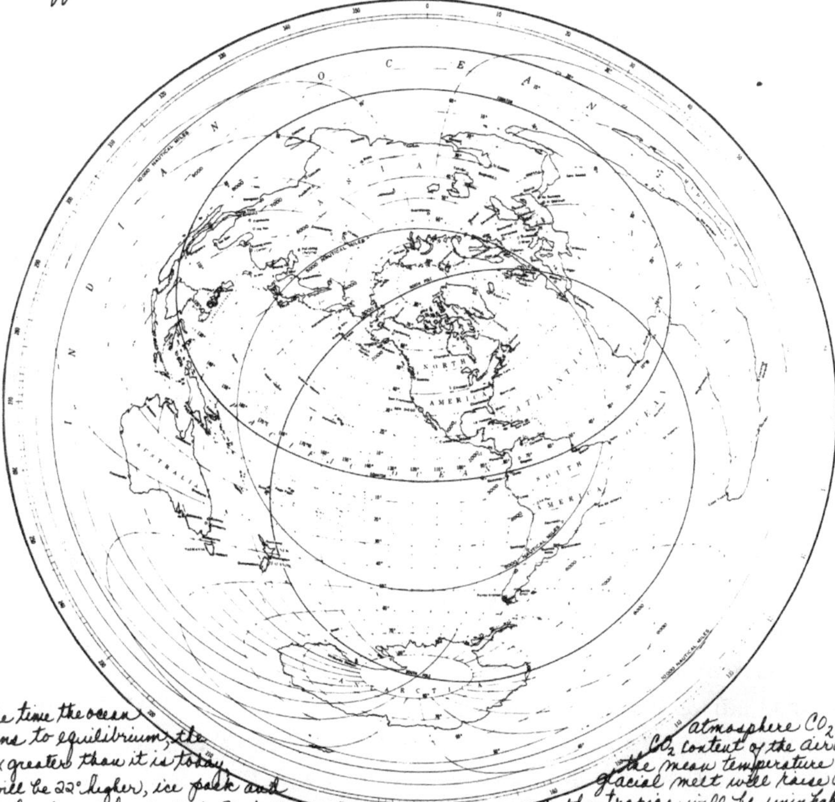

Then
by the time the ocean returns to equilibrium, the atmosphere CO_2 cycle CO_2 content of the air will be 10x greater than it is today the mean temperature of the air will be 22° higher, ice pack and glacial melt will raise the ocean levels at least 20 to 30 feet, many coastal areas and the tropics will be uninhabitable

Then
the land mass available will be X
the populations it can support will be X
the salinity of the oceans will be X
the mean temperature of the oceans will be X
the mean temperature of the air will be X
the resulting ecological transformations will be X
the resulting human social and political systems will be X

FIGURE 1.6 (continued)

Despite fidelity to the subjective and dramaturgical roots of scenario planning, Flohn has no hesitation in ascribing probabilities to the different futures under consideration (for example, in the case of global cooling, a very low probability of between 0.1 percent and 1 percent within the next century).[89] He also draws consistently on a combination of climate models and the more expansive sociopolitical models produced in the Club of Rome's *Limits to Growth* report. This amalgamation of models and scenarios represents a hybrid approach that survives to the present day in scientific literature on climate change, for example, when IPCC reports typically nest models within an overarching scenario architecture. As intuitive as this convergence might now seem, for many years the relationship between scenario planning and predictive modeling was a lot more fraught. According to his longstanding colleague Napier Collyns, Pierre Wack "regarded computer modeling as the enemy of thought," and many at the Hudson institute were not particularly keen either.[90]

Vocal opposition met the *Limits to Growth* report in the 1970s, criticizing its use of computer modeling—in particular, a systems dynamics computer model called World3, developed at MIT to simulate future trends. In 1980, a US-funded report titled *Global 2000* trod similar ground to the Club of Rome's publication, with both describing futures in which unbridled economic growth quickly ran up against planetary constraints.[91] Again there was vociferous neoconservative opposition, including a book-length response edited by Hermann Kahn and Simon Julian. One of the central arguments of *The Resourceful Earth: A Response to Global 2000*, which was published after Kahn's death in 1983, was a methodological critique of computer modeling, and in its introduction, Simon and Kahn quote approvingly Gunnar Myrdal's objections to the *Limits to Growth* report's reliance on a "world simulation model" that "represents quasilearnedness of a type that we have, for a long time, had too much of, not least in economics."[92]

If predictions made on the basis of computer modelling initiatives were the "enemy of thought," as Wack claimed, this is perhaps because they appeared to inhibit the ability of decision makers to "think the unthinkable"—not only in the sense of envisioning the untold miseries of a World War III, but in making imaginative leaps into seemingly improbable futures, in all their subjective and aesthetic dimensions. This tension between the capaciousness of scenario planning and the blinkered simulations of computer models is slowly resolved in climate research by synthesizing the two, not always to universal approval. From a critical futures studies perspective, the way in which the IPCC in particular has handled scenario planning can be criticized for its conservatism, evidenced by its only dealing with "policy-relevant" scenarios, and "excluding most futures as too utopian and dystopian."[93] Here, there is no place for the kinds of images that Polak, Boulding, and others saw as decisive for helping to shape desirable futures. Moreover, and despite its regularly issued warnings of ecological devastation if certain emissions targets aren't reached, for some, the IPCC's focus on the "most-likely" and "middle-of-the-road" scenarios strip scenario planning of a key feature that distinguishes it from prediction.[94] No longer tools for stretching the imagination of decision-makers or encouraging participation, in much scientific literature, scenarios are now objectified as "stand-alone scientific 'products.'"[95]

In this context, there are a number of artworks unblackboxing this product, taking it apart and reconnecting scenario planning to its creative roots. In *San Diego as the Center of a World*, the Harrisons took two predictions and transformed them into scenarios through a simple juxtaposition, without, for that matter, having direct knowledge of scenario planning as a distinct methodology.[96] Thirty-five years later, in a large project titled *Greenhouse Britain*, funded by the Department for Environment, Food and Rural Affairs, the Harrisons made use of scenario planning once again, this time explicitly. Considering the impact of climate change on the coastlines of Great Britain, the duo cast their gaze onto futures in which entire towns, cities, islands, and coastal communities are engulfed by the sea; millions of people are displaced; and plant and animal life retreats to higher ground. Although these are dramatic scenes, crucially, the project also emphasizes the gradual nature of these changes: "The waters will rise gracefully; how might one withdraw with equal grace."[97] In a case study focusing on the city of Bristol, a counterexample is given that demonstrates what an ungraceful response would look like: Dams and other environmental engineering projects designed to protect coastal towns, while successful in the short term, struggle against the inevitable, ongoing encroachment of the sea in the longer term.

What is significant about *Greenhouse Britain* is that, despite its meticulously researched content, the scenarios presented are not framed as objective products awaiting communication as polished, deliverable packages; from their inception through to their exhibition, they are participatory in nature. As the Harrisons note in their description of the piece:

> the idea was to democratize the information about what the water rises would do to the shoreline, so that each person could see what a two- or four-meter rising of the sea would do to the area where they lived. The implication was that every person facing this emergent event could become their own planner, or join groups to do so.[98]

Without departing from the realm of the possible entirely, the artists provide an epistemological backdrop from which participatory planning responses might emerge. Populating these futures of sea-level rise with responses of their own, one of the exhibitions that comprised *Greenhouse Britain* featured speculative designs for new buildings, landscape-scale settlement plans, and carbon sequestration strategies (figure 1.7).

Greenhouse Britain can be seen as part of a longer history of attempts to open up the practice of scenario planning to a more diverse and engaged usership beyond scientific and policy arenas, in continuity with an approach pioneered within the emancipatory tradition of futures studies. Several decades before, Galtung and others rallied against what they perceived as the elitism of mainstream futures work, populated by "high priests of the future."[99] More recently, the Harrisons' efforts can be aligned with initiatives embodied in the interdisciplinary research led by scholars such as Renata Tyszczuk, whose multifaceted work on the cultural dimensions of climate scenarios includes publications, programs, and partnerships underlining the need for collectively generated scenarios "fleshed out through collaborative practices of experimentation and improvisation."[100] This call is echoed by other social science and humanities scholars such as Kathryn Yusoff and Jennifer Gabrys, who have sought to highlight the role of the imagination in responses to climate change through a review of a handful of art projects that depart from the dominant framing of scenarios as authoritative, stand-alone products.[101]

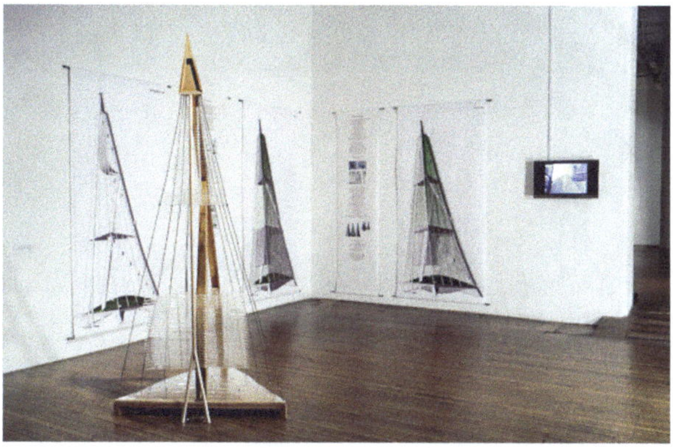

A recent film that takes up this mantle is by Lena Dobrowolska & Teo Ormond-Skeaping. Part of a larger project entitled *Future Scenarios* (2016–2020) that consists of photographic work, research documents, and text, the three-screen film installation uses footage that speaks of the myriad effects of climate change, from increases in the frequency and severity of disasters to diminished food security (figure 1.8).[102] This footage is punctuated by intertitles that contain a calendar moving forward from year to year, a colored dot fading from yellow to dark red, and a list of bleak statistics and possible events slowly appearing by their side, such as "95% of coral bleached" and "Disaster Preparedness taught in schools." Around twenty minutes before the end of the film, the year comes to a resting place at 2100, giving the viewer a clue as to the inspirations for some of the scenarios presented: Shell's Future Energy use scenarios and the United Nations Framework Convention on Climate Change's RCP scenarios, with which they are aligned. Both are products of scenario planning, and both use 2100 as a cut-off date for their speculations.

FIGURE 1.7

Helen and Newton Harrison, *Greenhouse Britain*, 2007–2009. Courtesy of the Helen and Newton Harrison Family Trust.

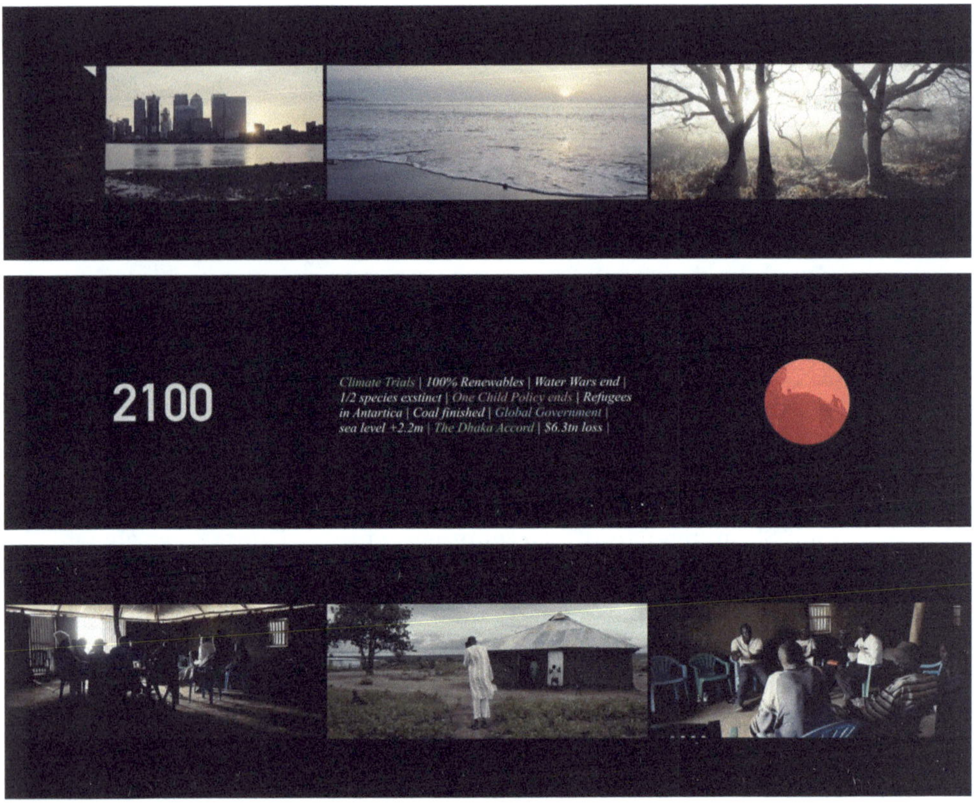

FIGURE 1.8

Lena Dobrowolska & Teo Ormond-Skeaping, stills from *Future Scenarios*, 2016–2020, three-channel installation, UHD, 2.35:1, 1 hour 10 minutes. Courtesy of the artists.

In addition to this textual component, the film makes visible the infrastructure of scenario planning through the spatiality of its installation. The multi-channel video installation, a common display strategy in contemporary art, here becomes a spatial metaphor for the relationship between present and futures. The footage on each of the three screens pulls the viewer into different environments and social situations, or variations on the same scene shown from different perspectives. While this multiplicity does not extend to the narratives the film explores—the three screens are not used to demarcate three discrete future scenarios—the division in perspective is nevertheless enough to create a visual analogue for the nonlinear dynamics of scenario planning, in which an array of futures branches away from the present along manifold pathways. It is not the narrative structure of scenarios that is on display directly, but the effects and affects of the scenario process—a dynamic not dissimilar to the way in which scholars in the energy humanities have discussed oil as something rarely seen or represented directly, but felt everywhere through the force of its effects.[103]

The positioning of the viewer as the point of division in this spatio-temporal arrangement has the secondary effect of attributing agency to the present—the first node in a decision tree from which scenarios start to branch out. This sense of agency may have been heightened further if the film didn't hew so closely to the 2100 cut-off date, which has been criticized for misrepresenting the scale of the challenge over the next few decades.[104] The effect of cumulative emissions means that the actions of humans during this period, a significant majority being the actions of affluent humans in Western countries, will have an impact for several thousand years to come. But while *Future Scenarios* misses this opportunity—perhaps necessarily, in order to foreground its link to the scenario planning projects of Shell and the UN—it does explore the multiple temporalities of the present in another way. In an interview, the artists point to the fact that for many in the West, and the UK in particular, "climate change remains a distant, far away, future that will largely happen to someone or something else in a landscape . . . most of us have never visited."[105] By using footage shot predominantly in the Global South, based in part on field research studying how Laos, Bangladesh, and Uganda are leading the way in their response to climate change, the filmmakers add a twist to the famous quote attributed to William Gibson that "the future has arrived—it's just not evenly distributed yet."[106] In this way, *Future Scenarios* complicates the narrative of a single future for a single planet, further shattering the fantasies of predictive futures work from the postwar era.

ACTIVIST ENERGIES IN MULTIPLE FUTURES

As has been shown, scenario planning cannot be restricted to a discrete method or technique belonging to a defined field of expertise. Whether through Watkins, the Metabolists, or environmental art, it has always, in one way or another, been implicated in an expanded field of speculative and creative practices. The management theorist and consultant Paul Schoemaker calls scenario planning a form of "disciplined imagination" because it helps train the minds of decision-makers on future possibilities they might not have previously considered.[107] But the disciplining of the imagination is an equally fitting term for the ways in which the soft edges of cultural production are sharpened and transformed into disciplinary borders, and the associated methods cocooned within the worlds of corporate, scientific, and public policy research against a backdrop in which futures work becomes increasingly fragmented.[108]

Part of the concern here is for what gets excluded (knowingly or otherwise) when a set of cultural practices is formalized into a methodology associated with a specific profession or area of activity. If business studies literature is to be believed, the infrastructure of scenario planning originates with Herman Kahn, extends outwards through corporate entities including Shell, before eventually finding a home in organizations such as the International Monetary Fund (IMF), the World Economic Forum, and the IPCC. This chapter has already sketched a broader context for this history, in which scenario planning draws inspiration from theatre, film, and science fiction; plays an active role in projects and organizations pursuing social and environmental justice; and comes to influence a range of art projects from the 1970s to the present day. In this final section, the historical brackets will be widened even further to consider a multifaceted art project more closely aligned with Afrofuturism. In contrast to some of the cultural artifacts above, the multimedia evolution of Afrofuturism neither owes a debt of influence to scenario planning nor could be said to have influenced its development in any direct fashion. The two share no common point of origin, follow very different trajectories, and on the face of it only seem to be united through their mutual preoccupation with futures-oriented thinking. For some, if there is a direct link between scenario planning and Afrofuturism, it is one characterized by antagonism rather than mutual reinforcement.

This antagonism surfaces in an influential essay on Afrofuturism by the theorist and filmmaker Kodwo Eshun, who groups scenario thinking together with a number of other modes of speculation within what he calls the "futures industry."[109] Driven primarily by market forces, the futures industry encompasses, for Eshun, "the intersecting industries of technoscience, fictional media, technological projection, and market prediction," which together objectify the futures of African and Afrodiasporic peoples. One of Afrofuturism's first tasks, for Eshun, is to expose the operations of the futures industry and critique scenarios and market projections that trade on the image of Africa as the site of countless future crises.[110] Framed in a discourse of "unchallengeable certainty," these futures act to "forecast and fix African dystopia."[111] In a discussion that is part description, part prescription, Eshun characterizes Afrofuturism as "a program for recovering the histories of counter-futures created in a century hostile to Afro-diasporic projection and as a space within which the critical work of manufacturing tools capable of intervention within the current political dispensation may be undertaken."[112]

Weaving together a number of different arguments, this quote simultaneously speaks of nonlinear temporalities, the politics of erasure, and the role that infrastructures of speculation can have in shaping present political realities. On this last point, there is perhaps more common ground shared by Afrofuturism and scenario thinking than it may at first appear. Both seek to make interventions in the chronopolitical realm, albeit often to very different ends, and both display a core commitment to elaborating futures in the plural rather than singular. While it is undeniable that scenario thinking has been an influential instrument of state and corporate planning, punctuating this history is a series of experiments with scenarios that are more creative, participative, and activist in their sensibility. Here, it is the very openness of the scenario technique that allowed activists to counter state and corporate interests.

What Eshun calls the "futures industry" could be allied to what Ahlqvist and Rhisiart identify as the "utilitarian" trajectory within futures studies rooted in the Cold War state apparatus, with its entrenched ideas of modernization that placed nation-states on a predestined developmental path to the future.[113] For such futurists of the post–World War II era, Western socioeconomic models served as templates that could be followed by countries in the "developing" world, often to ruinous results. A significant challenge to such ideas was posed by global events in the 1960s, as well as by the scholars and activists who coalesced in the WFSF.[114] While many of the leading figures in the early days of the organization were based in Europe and America, their stated ambition to unleash "indigenous, alternative, and dormant future models for the developing world" eventually contributed to a more global movement.[115] Long-range studies explored alternative development pathways in north African countries such as Egypt, where a group of futures researchers gathered around Ibrahim Helmi Abdel-Rahman, and in Morrocco, where Mahdi Elmandjra held the WFSF presidency from 1977 to 1981.[116] The influential Indian scholar Ashis Nandy would develop futures work from within the Centre for the Study of Developing Societies, Delhi, and later on, futures work would undergo what Wendy Schultz describes as a "pacific shift," whereby many of its key research centers, training programs, and journals would be established in places such as Hawaii, Australia, Taiwan, Singapore, and South Korea.[117] In contrast to narrow discourses of modernization propagated by think tanks and research institutes in the West, many of these initiatives used scenario planning to pluralize the modes of development on offer, proliferating the number of "images" of futures available.

Returning to Afrofuturism, it's possible to see a similar thirst for alternative futures in many of the artists, authors, and musicians that identify with the designation. The group Black Quantum Futurism (BQF), comprised of Camae Ayewa and Rasheedah Phillips, is an exemplary case. In multimedia projects that include music and spoken word performances, participatory workshops, innovative forms of archiving, and interactive installations, the group actively challenges what its members see as the oppressive nature of Western, linear clock time. Drawing on an expansive body of research, from anthropological accounts of African concepts of time to postcolonial critiques of the "settler time" that forcefully displaced them, BQF develop novel methodologies that treat time as something that is always local rather than universal.[118]

Through organizing workshops on "community futurisms" using such devices as "quantum event maps" and "oral futures booths," a number of BQF's projects embody the participatory intent of some scenario planning, but, crucially, they go beyond it. As well as treating futures as multiple, contingent, and impossible to fully predict, they extend such a disposition toward the past, introducing an important temporal complication that amounts to what the writer Sofia Samatar calls an "entanglement of the tenses."[119] In Afrofuturist thought, this often entails reclaiming counter-histories of oppression and emancipation to support ongoing struggles, so that history becomes a function of present interpretation and the futures it seeks to support. However, BQF's quantum event maps also complicate the very means by which time is organized by playing with the linear axis that proceeds from past to present to future, something it shares with Indigenous Futurism's concept of the "native slipstream," in which "pasts, presents, and futures . . . flow together like currents in a navigable stream."[120]

One context in which BQF has used quantum event maps is in its members' hometown of Philadelphia, and specifically in the city's Sharswood neighborhood. Here, a tidal wave of redevelopment and gentrification in recent years has resulted in the displacement of a historically Black population, with many homes bought through eminent domain (a process sometimes called "expropriation" or "compulsory purchase" outside of North America). As a public interest attorney specializing in housing law as well as an artist, Phillips brought a wealth of practical expertise to a project that was nevertheless avowedly speculative in its intent. Titled *Community Futurisms: Time & Memory in North Philly*, the project established a community space where Afrofuturism was used as a concrete strategy for "dialogue, visioning, and testing of ideas around community sustainability, resilience, and resistance—and as a technology for the actual implementation of those visions and ideas."[121] This represents a powerful cognitive antidote to the kinds of top-down futures work sometimes promoted through centralized planning initiatives, as well as the sensory yet somewhat illustrative science communication of *Climate Garden*. The project also acts as a foil to a certain strand of Afrofuturism that operates in a more restricted arena. As Phillips has made clear, "Afrofuturism is often pigeonholed or stereotyped as being primarily concerned with Black science fiction, or entertainment or aesthetics. . . . It should increasingly be applied to other areas of study, such as law, architecture, psychology, social work, public policy, etc."[122]

As their intervention into Sharswood shows, many of BQF's projects seek to develop an Afrofuturistic social practice that intervenes at the level of physical and social infrastructure. In this case, a key target was the planning infrastructure that ensures Black communities are locked out of decision-making processes about their own housing provision. And yet, when processes of timekeeping are understood as infrastructures in and of themselves, the area of application can be viewed as wider still. The linear timelines that pass from the present to the future, on which redevelopment of low-income neighborhoods is positioned as inevitable, are like train tracks that lock in certain developmental logics and exclude others. While only one part of a larger constellation of tools, the specific role of quantum event maps here is to intervene at this level of temporal infrastructure, opening up ways of constructing time differently, and in particular those ways that mimic African and Asian cultural practices that pull events around a central focal point, rather than plot them onto a straight line or a causal chain. While devices such as quantum event maps are not in themselves an attempt to represent specific pasts, presents, and futures, they nevertheless prompt those who use them to engage in a highly aesthetic mode of visioning. Events are addressed in all their sensuous multiplicity, broken down into such categories as smells, colors, sounds, and moods, demonstrating what an expanded field of speculative thinking can include beyond the routine deployment of futures methods in corporate and policy arenas.

While not immediately concerned with the environmental futures that sit at the core of this book, many of BQF's interventions operate at the level of temporal infrastructure, meaning that the resulting tools and techniques have the ability to move between areas of application. The context of the quantum event maps discussed above is a project about the spatial injustice suffered by Black communities in a specific place, and yet, it is easy to imagine the same tools operating in relation to struggles for environmental justice, echoing the numerous voices that have highlighted the connection between environmental degradation and racial oppression as mutually reinforcing historical processes.[123] This form of cross-disciplinary drift is actively encouraged by Phillips, who speaks of the need for "Afrofuturistic Environmental Science, Afrofuturistic City Planning, Afrofuturistic Medicines, Afrofuturistic Math, Afrofuturistic Horology."[124] In the formation of these areas of activity, and practices of researching them historically, a powerful argument for an undisciplined aesthetic expansion of scenario-based thinking comes into view, one that breaks free from a circumscribed set of competencies and the conservatism of IPCC pathways and explores the manifold entanglements of counter-histories and counter-futures toward emancipatory ends. It is in examples such as these that scenario thinking encounters more expansive practices and traditions. In the similar way that Kahn and others cultivated techniques that would stretch the predictive imagination into multiple possibilities, today, a whole range of futurisms push this multiplicity further, exploring the plasticity of different temporal registers through embodied aesthetic practices in ways that have the capacity to provide new "infrastructures of feeling" better aligned to the chronopolitical complexity of environmental crisis.

2

PROSPECTING

Just at the time when people are despairing at realizing that they might, in the end, have "no future," we suddenly have many prospects.
—Bruno Latour, "An Attempt at a 'Compositionist Manifesto'"[1]

INTRODUCTION

In the spring of 1867, the geologist Clarence King embarked upon the first of several ambitious surveys of the American West. Charged with documenting that "belt of country" that ran along the fortieth parallel from Colorado to Nevada, King took with him a team of nine specialists that included a botanist, a topographer, a zoologist, and a photographer named Timothy O'Sullivan.[2] While the group's expertise was mostly scientific, the aims of the expedition were heavily influenced by commercial factors, with a focus on the productive capacity of the landscape and its untapped reserves of silver and coal. In this respect, it is telling that, of the seven volumes of King's magnum opus published after his survey expeditions, the first to appear was not volume one, but volume three, simply titled *Mining Industry*.[3]

Many of the photographs O'Sullivan took on these surveys have since acquired the status of highly prized aesthetic objects, their depictions of magisterial canyons and ancient geological formations attracting the

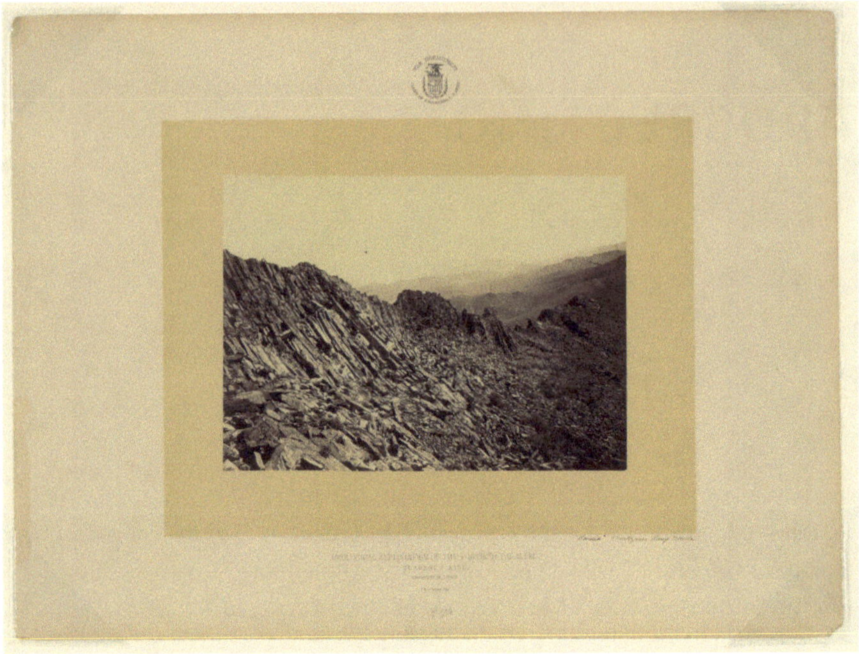

attention of subsequent photographers such as Ansel Adams. And yet, however beautifully images such as figure 2.1 captured the rugged terrain of the American West, they also went hand in hand with the aims of economic and colonial exploitation and provided a "prospect" from which to view the future. As well as possessing undeniable aesthetic merits, O'Sullivan's photographs are fine-tuned to portray a frontier charged with the promise of untapped potential, and many of the photographs he

FIGURE 2.1

Timothy O'Sullivan, *Karnak, Montezuma Range, Nevada*, 1867. Plate 23 from Clarence King et al., *Report of the Geological Exploration of the Fortieth Parallel*, vol. 2, *Descriptive Geology* (Professional Papers of the Engineer Department, U.S. Army, 1877), 763.

produced between 1867 and 1874 were of direct benefit to future mining and military endeavors.[4] For this reason, they can be viewed as early examples of what could be called "visual prospecting," a form of image-making that enables, expedites, and sometimes critiques the pursuit of profits by means of extraction.

O'Sullivan's survey photographs are not the primary focus of this chapter, but rather a springboard that launches the inquiry into a discussion of the visual culture of prospecting today. Despite the temporal disjuncture, these faded photographs of the American West are nevertheless connected to more contemporary images of extraction in a number of ways. O'Sullivan's work provides more than a historical backdrop to this contemporary work; it helps to demonstrate that while technologies and techniques may have changed, many of the drivers and underlying representational dynamics of visual prospecting have remained remarkably constant over the past 150 years. It also serves to reveal an important feature of prospecting: that it is rarely the preserve of individuals or individual disciplines working alone. As surveys like Clarence King's demonstrate, prospecting more often involves large, coordinated efforts in the search for resources, which are themselves bound up in wider infrastructures of knowledge production of which vision and visualization are key components.

In contrast to the speculative infrastructure of scenario planning discussed in the last chapter, which has a clear, if complex, origin story that can be traced through a series of shifts that took place in the middle of the twentieth century, prospecting has no unambiguous point of origin, no progenitor, and very few definitions that take account of the different forms of extraction it entails. In its link to mining, prospecting is situated at the point where geological knowledge and economic motivations coincide.[5] In more recent decades, however, the word has also come to be used in a more directed sense, principally in practices of "bioprospecting," which involve the search for biological and genetic resources, almost always through forms of dispossession.[6] Whether the material in question is obtained by clawing minerals from the ground or from mining so-called "traditional" ecological knowledge, in each case, the process sits somewhere between extraction and speculation, constituting an informed guess on where profit might be gained in decades to come.

In this respect, prospecting is characterized by its persistent attempts to look ahead, underpinned by an epistemology that reduces the world to a basket of undiscovered resources in the present. In recent decades, prospecting has been turbo-charged by climate change, as governments and corporations compete to secure the minerals necessary to support decarbonization, and seed companies seek out new varieties able to thrive under changing conditions.

As a noun, a prospect is also an inescapably visual concept, closely linked to practices of viewing a landscape that enable an encompassing overview, or an extension of the typical range in which vision operates. This visual orientation is something it shares with speculation, to the extent that both words are in part derived from the Proto-Indo-European root *spek*, meaning "to observe."[7] In analyzing the visual culture of prospecting, this chapter stays within the orbit of this meaning, and by attending to the diverse ways in which prospecting manifests, it offers something of a panoramic view that resists the temptation to narrow its focus on any one material or visual technique. While the next section attends to the historical association between prospecting and the view from above, the diversity of examples that follow is intended to complicate the idea that prospecting necessarily requires an overview, or that the view from above is always a harbinger of resource extraction to come. By shuttling back and forth between a broader visual culture of prospecting and specific works of art by the likes of Edith Morales and the artist group On-Trade-Off, the aim is to loosen the association between the overview and what has been called the "extractive view," highlighting the diversity of contexts in which the visual culture of prospecting manifests.[8] O'Sullivan's photographs provide a useful first step in this direction, since they are images that facilitated resource extraction well before the aerial view had become the ubiquitous phenomenon it is today in the age of satellites. The selection of examples in what follows further pluralizes the extractive view, whether through representations of the subsurface of the earth, images designed to attract speculative capital, or depictions that isolate representative "specimens" by means of standardized visual conventions. In doing so, it points to multiple ways in which the speculative infrastructure of prospecting is made visible against a backdrop of unprecedented environmental change.

EXTRACTIVE VIEWS

From the high ground prized by generals on medieval battlefields to the strategic advantage afforded by access to satellite technologies in the twentieth century, there is a link between the aerial view and forms of power and domination that runs deeper than any parallel development. Despite its frequent claims to objectivity and neutrality, the aerial view—what Donna Haraway influentially called "the view from nowhere"—has acted as both a perspective on the world and a powerful instrument for changing it.[9] The literature on the topic is large and diverse, and often frames the issue in terms associated with military and state power.[10] There is also a sizable body of work that looks at the role of the aerial view in the reconfiguration of land and ecosystems as inventories of resources, a process intimately attached to colonial violence.[11] A key reference here is James C. Scott's discussion of attempts to establish a scientific approach to everything from forest management to urban planning, which first of all required the ability to "see like a state."[12] On another track, it was only by means of vast quantities of state funding that the famous blue marble images taken from the Apollo XVII spacecraft would come to energize the environmental movement and shape countercultural initiatives such as the *Whole Earth Catalog* for decades to come (chapter 3 will address this more "macroscopic" view from the perspective of early ecological science).[13] It is largely by means of these tributaries that discussions of the aerial view enter the environmental arts and humanities, where it is couched in such various terms as a "God's trick," "premonitory perspective," or "extractive view."[14]

To take two recent examples, in his analysis of the large-scale aerial photographs of Edward Burtynsky and Louis Helbig, T. J. Demos speaks of the complicity of the artists' work in "dramatizing in spectacular fashion the perverse visual beauty of a technological, and even geological, act of mastery devoid of environmental ethics."[15] Here, Demos suggests that the aerial perspective (albeit often an oblique one in Burtynsky's work) obscures the wider socioeconomic effects of extraction, presenting a depopulated resource frontier seemingly unconnected to either the local or global economy. Aerial photographs of the California oil fields or the Alberta tar sands are explicitly contrasted with a photo exhibition titled

Petrochemical America by Richard Misrach, which shows "on-the-ground environmental and human costs."[16] Whereas, for Demos, the aerial view and remote sensing tend to "fetishize mastery of the visual field," there is an implication that this tendency can be undercut by images, like Misrach's, that show the ravages of extraction from a more terrestrial perspective.[17]

In a similar vein, for Macarena Gómez-Barris, it is by focusing on "extractive capitalism and its vertical model of seeing" that we can better understand the ways in which colonial and state violence is strengthened through visual regimes. Drawing on colonial and environmental histories, particularly in Latin America, Gómez-Barris repeatedly associates "vertical seeing" with "colonial visual regimes that normalized an extractive planetary view that continues to facilitate capitalist expansion, especially upon resource-rich Indigenous territories."[18] Whether by means of historical examples of extraction and displacement or newer forms of satellite imagery that facilitate high-resolution resource mapping, the vertical and the extractive are often spoken of interchangeably, and both are opposed to what Gómez-Barris calls the "submerged perspective," as more enmeshed in the material encounters of the world.

Although Demos's analysis is rooted in a discussion of specific artists, and Gómez-Barris's use of the word "vertical" doesn't always seem to map onto actual visual cultures, their contributions nevertheless help reinforce a wider binary between the vertical perspective as objectifying, and the horizonal as a partial perspective in which political and environmental ethics inhere. While it is undeniable that the aerial view has facilitated the very worst forms of extraction and colonial violence, as Heather Houser points out, "the view from above is not proprietary to any one epistemological stance."[19] Houser's problematization of recent discussions of the aerial view works by placing it in specific technological contexts, emphasizing processes of techno-mediation and how these become the subject of artistic and activist reflection.[20] What follows in this chapter draws inspiration from this approach, but instead of focusing on technology per se, it considers the importance of wider infrastructures of knowledge production and the socioeconomic contexts in which they operate. In doing so, it seeks to provide a more holistic consideration of visual cultures of extraction and their connection to environmental pasts and futures, rather than viewing them exclusively through a binary framework or emphasizing the technical means of image production.

RARE EARTH

From the abandoned nineteenth-century mining towns of Nevada photographed by O'Sullivan to the flooded pocks and holes that riddle the landscape surrounding Manono in the Democratic Republic of Congo, the places where prospecting leads to economic boom often go bust just as quickly. In contrast to some of the towns that O'Sullivan photographed in the nineteenth century, the promises of riches in Manono have not yet been exhausted. After a crash in the price of tin in the 1980s put an end to extraction that had continued more or less uninterrupted since 1915, Manono is now the site of significant speculative investment targeting its vast lithium deposits—a raw material whose price has skyrocketed as it assumes the status of a critical raw material for many nations as they undergo green energy transitions.[21]

The cycle of boom and bust in extractive economies such as Manono's is the subject of a number of images produced by the Congolese photographer Georges Senga, which feature in an exhibition by On-Trade-Off.[22] In three complementary series, *Tshanga-Tshanga (300 Maisons)* (figure 2.2), *Tshanga-Tshanga (Barzin)*, and *Tshanga-Tshanga (Mille Bêches)* (figure 2.3, all 2022), the artist uses the aerial view to document the decaying remnants of the last century's extractive economy. In the first series, his drone-mounted camera trains its sights on the once comfortable housing infrastructure constructed for managerial staff at the company, contrasting this with smaller working-class homes photographed from street level in another series, *Tshanga-Tshanga (Barzin)*, named after the Belgian architect and engineer Henry Barzin (1884–1971), who designed many of the buildings in the once profitable colonial mining outpost.[23]

The word 'tshanga' [or changanya] is commonly used in Swahili in the sense of mixing or blending, but it also carries a more complex set of associations linked to the practice of creating patchwork quilts from scraps of leftover fabric. In *300 Maisons*, an overhead view turns the buildings into a patchwork of different rooms. The large concrete and brick villas

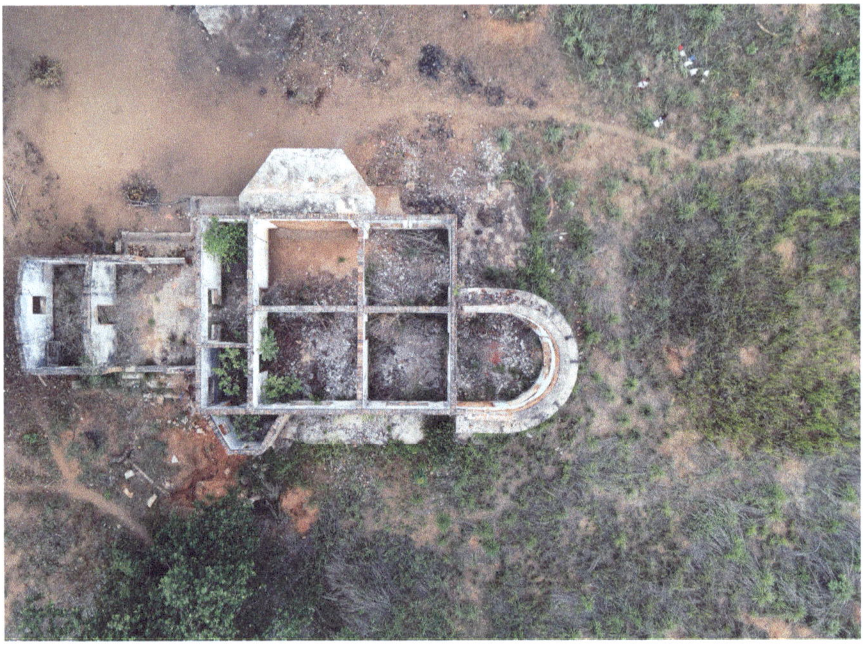

lie in a state of abandonment. Roofs have fallen prey to time and violence inflicted during the civil war in the late 1990s, so that room divisions are exposed to the aerial view, reminiscent of the architectural drawings used to sell buildings off-plan. In this way, the photographer's treatment of the ruins simultaneously signals forward and backward in time, layering future promises of development on the remnants of a colonial past. *Tshanga-Tshanga* also refers to an ethos of ethnic mixing, which was paternalistically mobilized by mining companies in the DRC to dislocate clan and filial bonds, and, as Senga explained to me in an interview, "to create the best worker."[24] In contrast to the *300 Maisons*, the working-class

FIGURE 2.2

Georges Senga, *Tshanga-Tshanga (300 Maisons)*, 2022. Courtesy of the artist.

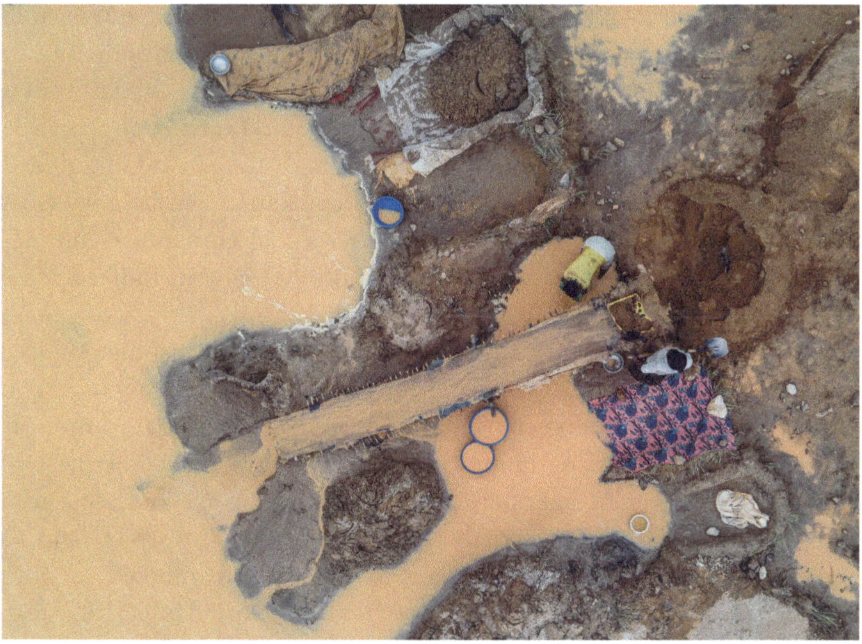

homes in the *Barzin* series are photographed from a frontal perspective, their uniformity testament to this atomistic approach, which is further emphasized by a grid-formatted display.

The third series, *Tshanga-Tshanga (Mille Bêches)* (figure 2.3), focuses on mineral extraction directly, again using the aerial view to document the efforts of the local population in their autonomous mining activities. Despite being photographed from above, these individuals are well above the "threshold of detectability" that limits the maximum resolution of satellite imagery in the interests of protecting privacy and national security.[25] And yet, the aerial perspective ensures they are not presented as

FIGURE 2.3

Georges Senga, *Tshanga-Tshanga (Mille Bêches)*, 2022. Courtesy of the artist.

individuals, only as a few of the thousand shovels (*mille bêches*) used to dig the sand quarry that still provides a modest income to the inhabitants of Manono in their search for cassiterite and tantalum, the latter now classified in many countries as a strategic metal. Much of this activity also takes place around the flooded leftovers of the old mining pits. A small sluice box serves as evidence of the continued quest for valuable minerals even after the withdrawal of the Belgian company Géomines in the 1980s—efforts at salvage accumulation that now seem likely to be replaced by more intensive forms of extraction.[26]

As these examples show, the selective use of an aerial drone-mounted camera in Senga's photographs derive a great deal of meaning from context, and the vertical perspective helps introduce a temporal complexity that resists the totalizing qualities of the extractive view. For the Uruguayan political ecologist Eduardo Gudynas, the kind of subsistence mining shown in *Mille Bêches* does not qualify to be called "extractivism," a term he reserves for a "high-volume and high-intensity modality of natural resource appropriation, where resources are exported in primary commodity form with little or no processing."[27] The current plans to kickstart lithium extraction in the region operate on an entirely different scale, with the Australian company initially spearheading the project, AVZ Minerals, intending to export as much as eighty thousand tons of lithium concentrate per month to battery manufacturers worldwide, although conflict in the country and efforts on the part of the DRC government to prohibit the export of unprocessed concentrates may complicate matters in future years.[28] Coupled with the fact that the mine is unlikely to create many jobs in the region, it looks increasingly destined to operate in what Fernando Henrique Cardoso and Enzo Faletto term an "enclave economy," insofar as its links to the global economy will be significantly stronger than those to the national economy of the DRC.[29]

When these photographs were made, Senga was based 600 kilometers south of Manono in Lubumbashi, a city with its own history as an enclave economy, having been established as a copper-mining settlement by the Belgian colonial regime in 1910. In recent years, Senga too has been involved in a transnational partnership around lithium. On-Trade-Off was initiated by the Congolese arts organization Picha together with the Brussels-based artist collective Enough Room for Space in 2018, when rumors about the vast lithium deposits thought to exist in Manono had already begun to circulate. The group has made lithium extraction its core concern, connecting this to a wider set of issues including energy futures, financial speculation, and neocolonial power relations. In contrast to the unequal exchange characteristic of enclave economies, the group has consciously pursued forms of collaboration that equitably share knowledge and resources, constituting something of a shadow knowledge infrastructure, albeit one operating at a very different scale to the infrastructures of extraction on which it nevertheless depends. For example, a growing collection of research materials circulate on an open-source platform and have provided raw materials for some of the artwork produced by the group, and a percentage of the profits from any sales feeds back into a collective fund. International travel is also used as an opportunity to exchange physical objects such as books and materials for art production that might be difficult to source in the artists' home countries.[30] An awareness of the uneven access to digital infrastructures is evident in the group's carefully designed website. With internet access in the DRC still expensive, the design uses images composed of ASCII characters as default, which require significantly less data than JPEG or PNG formats. Other design features include a banner stating the amount of CO_2 produced with each visit to the website, and the consistent use of black backgrounds, which require less power to display than brighter colors.

In the relatively short life span of the group, the members of On-Trade-Off have used these platforms to make and promote artwork both individually and collectively, give presentations, and participate in exhibitions and residencies. *Charging Myths* was their largest exhibition at the time it opened, occupying two floors of the Z33 House for Contemporary Art, Design & Architecture, in Hasselt, Belgium, and featuring visually diverse contributions yoked together through a shared preoccupation with lithium and electricity.[31] The aim in including a discussion of the exhibition in this chapter is not to provide a critical overview or evaluate its successes or failures. Rather, the intention is to demonstrate that, like O'Sullivan before him, Senga's photographs are bound up in wider infrastructures of knowledge production, of which the exhibition is but one public-facing outcome. The consideration of these infrastructures and their socioeconomic context compels viewers to think through the

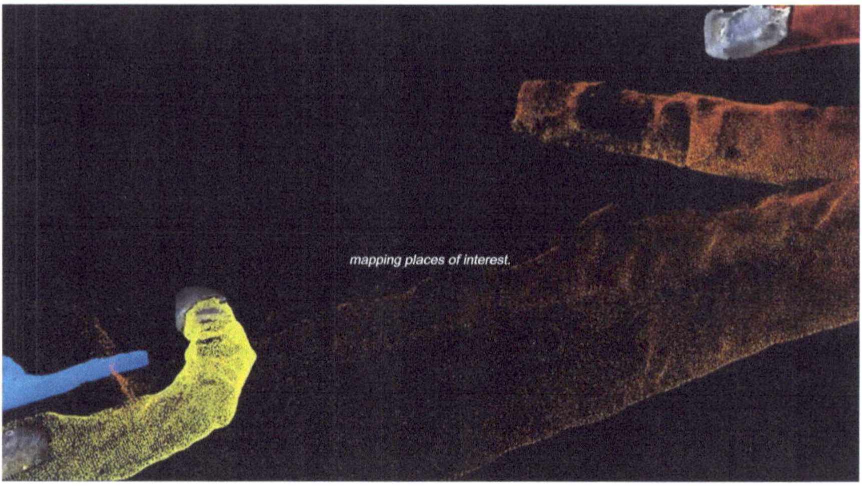

nuances of how prospecting operates, and makes it difficult to label one particular visual form such as the aerial view "extractive" and another "grounded" or "submerged" without first giving consideration to the exchanges of knowledge and resources that shape their creation and dissemination in the world.

Femke Herregraven's *Prelude to: When the Dust Unsettles* (2022) is a work that takes a different approach in its exploration of visual prospecting, blending footage of the site of the proposed lithium mine in Manono with digital promotional videos produced by mining companies that demonstrate cutting-edge technologies of extraction. A dizzying array of data visualizations present a landscape visible from all angles, where solid bedrock is rendered transparent at the click of a button. A garish RGB color palette assembles cloud point data into ghostly snaking forms that trace the contours of a mine's shafts and tunnels (figure 2.4). The

FIGURE 2.4

Femke Herregraven, *Prelude to: When the Dust Unsettles*, 2022, 15:25 min., video still. Courtesy of the artist.

camera footage accompanying this computer-generated imagery is at times just as disorientating, captured by a variety of drones and robots as they quietly probe subterranean passages devoid of human life.

A written on-screen dialogue elegantly threads its way through the film, staging a conversation between the mine as it currently stands and its digital twin existing as an idealized projection. "They survey me as if I was you," remarks the digital twin, while the mine is keen to point out differences: "Where is the ecosystem?" "How can you be my twin when you reduce me to a single mineral?" This dialogue loosely maps onto the two different types of footage in the film, becoming more pronounced toward the end as the viewer is led above ground, into a landscape that feels at first like it belongs in a first-person computer role-play game, before this is intercut with images of the dusty red soil of Manono itself, complete with evidence of low-tech artisanal mining activity. Accentuating the division is a soundtrack composed by B. J. Nilsen that uses audio samples of artisan miners digging with hand tools in Manono recorded by another member of the collective, Gulda El Magambo.

Below ground, it is harder to draw a neat separation between the two twins, as many of the tools used to gather film footage seem also to be scanning the environment to gather enough data to create a digital facsimile. The points of passage between the two types of imagery are explored in another short section, in which one of the few humans to feature in the artwork is filmed in an immersive "virtual blast wall" environment, where physical tools are rigged up to motion capture systems, enabling movements to be translated into the virtual environment in real time.[32] The environment is the creation of the South African company Simulated Training Solutions (or STS3D), who have developed the use of virtual reality to equip miners with the skills necessary to carry out dangerous operations underground, such as preparing blast charges.

The footage culled from STS3D's promotional videos dazzles with its promises of a clean, safe, and ultra-modern mining environment, quite unlike the extraction depicted in some of Georges Senga's photographs. Indeed, this allure is something the promotional source material for Herregraven's video seems to deliberately cultivate. Much like the slick marketing videos and fly-through animations that help to sell luxury properties to investors, an important aspect of the relationship between visual culture and the extractive industries concerns the production of expectations. As researchers in a range of fields and locations have shown, from genomics in Iceland to gold mining in Indonesia, expectations are deliberately cultivated as a means to attract investment, broker relationships, and set out agendas.[33] Even when expectations are not formalized into contracts or promissory statements, they nevertheless have performative effects, creating path dependencies based on shared narratives.

In the case of mineral extraction, the production of expectations often involves a series of identifiable visual tropes. The proposed mining site is typically shown devoid of all human or animal life, creating an imaginary resource frontier of "wild and empty spaces where *discovering* resources, not stealing them, is possible."[34] While the source material of Herregraven's video could not be more different from O'Sullivan's survey photographs in production method or aesthetic, they nevertheless both show a depopulated landscape, whether through digital or analog means. As has been noted in relation to the King survey, there was little expertise in the human sciences of archaeology and ethnography among the exploration team, and evidence of Indigenous populations was conspicuously absent from the resulting visual and written material.[35] Likewise, with the exception of those demonstrating STS3D's virtual blast wall, the source material of Herregraven's video installation emphasizes the topographical and geological properties of the proposed mine site, creating a "digital twin" in which human and animal life is almost entirely absent.

With their "thresholds of detectability," aerial views are certainly one way to depopulate a landscape, and are often bound up in the production of expectations that fuel speculative accumulation in places like Manono. Aerial views also play a small but significant role in *Prelude to: When the Dust Unsettles*, insofar as the artwork has been installed using a ceiling-mounted projector beaming the film onto a contoured surface on the floor of the gallery, forcing the viewer to adopt an elevated view when they encounter the work. Floor projection aside, the diversity of tools and technologies of visualization at work in the video demonstrates that contemporary prospecting can function perfectly well without the kind of overhead view that has historically facilitated various forms of extraction.[36] Both Senga's photographs and Herregraven's video demonstrate the plurality of extractive views at play in mineral prospecting, wherein the dichotomy is not simply between vertical and horizontal axes of vision but between landscapes as products of layered histories and multispecies relations and landscapes as the bearers of future desires focused on a handful of lucrative resources. As complementary perspectives on the same physical location, Senga's photographs show how the aerial view can unveil complex social forces beyond extraction, while Herregraven's video reveals how the visual culture of contemporary prospecting is more heterogeneous than existing scholarship might suggest. As two outcomes of the collective project On-Trade-Off, they demonstrate how such artist-generated knowledge infrastructures have the capacity to both mirror and undermine the production of visual culture that supports extractive endeavors.

GREEN GOLD

The establishing shot of Edith Morales's short film *Aerial Root* (2020) is slow to appear. Thin clouds drift past the lens of the drone-mounted camera as it makes a gradual descent to the lush vegetation below. The camera eventually finds itself at ground level amidst a field of maize and a soundscape of loud birdsong. Viewers whose experience of agriculture is limited to industrial monocrop farming may not immediately recognize the landscape as one of food production, with irregularly shaped fields sitting between clusters of large flowering trees and no obvious boundary markers in sight. When the camera trains its sights on the leaves and tassels of several maize plants, it is clear that this is a landscape that is as productive as it is picturesque. As the film's voiceover explains, the land is being cultivated using the *milpa* farming system, one aspect of which involves a single field being planted with several complementary crops—typically a mixture of beans, squash, and maize—that have adapted to the region over thousands of years. While the squash provides ground cover to help with moisture retention, the beans deposit nitrogen into soils that would otherwise struggle to satisfy a typical maize variety's appetite for nutrients. Originating in the artist's native Mexico/Mesoamerica, the milpa system can now be found across the world in a range of alternative approaches to agriculture, where it joins other multicrop systems as a key component of more recent practices such as permaculture.[37]

The star of Morales's short film is a particular kind of maize known as *olotón* that, together with a handful of other varieties, possesses the rare ability to synthesize its own nutrients by converting atmospheric nitrogen into ammonia.[38] As the camera brings the viewer into ever closer contact with the plant, the focus becomes a series of short brace roots that glisten with a strangely gelatinous mucus that carries out this feat with nitrogen-fixing bacteria. A particularly useful crop as soil degradation is exacerbated due to climate change, olotón maize has now attracted the attention of large research universities and multinational food giants such as University of California, Davis and BioN2, a subsidiary of Mars Inc.[39] As the voiceover to the video explains, this has led to an attempt to patent the genetic characteristics of the plant, with a view to cross-breed it with other commercial corn varieties, "forcing it to enter the circuits of global distribution."[40]

The film is keen to contrast the extractive endeavors of these companies with the socio-ecological context from which olotón maize has emerged over countless generations. As well as focusing on the corn itself, the camera captures the communities that cultivate it, with hands tending to the young plants, and their eventual yield displayed by Mixtec women in front of their houses. The voiceover further reinforces this narrative, putting forward a robust critique of organizations such as University of California, Davis (UCD) and BioN2, whose joint patent application it describes as "biopiracy" and "an act of pure colonialism."

Another term critics sometimes use interchangeably with "biopiracy" is "bioprospecting," which denotes the search for genetic, chemical, and biomechanical properties of living things that have commercial potential.[41] In addition to contributing to the R&D initiatives of the agricultural sector, bioprospecting has become a controversial topic in pharmaceutical research, with numerous patent applications attempting to codify Indigenous knowledge as the intellectual property of large corporations. One of the key requirements of any patent application is to demonstrate the novelty of the product or process under review, a feature that the anthropologist Cori Hayden attributes to the legacy of John Locke's Enlightenment notion that benefits should be derived from "the fruits of one's labor."[42] This requirement notwithstanding, there has been a slew of patent applications in recent years. A list of high-profile examples might include the attempt by the California-based entrepreneur Loren Miller to patent a vine used to produce the hallucinogenic drink ayahuasca; numerous patent applications relating to the neem tree, native to the Indian subcontinent; or proprietary claims on the cactus *Hoodia gordonii*, native to southern Africa, where it has been used for generations by the San people of the Namib desert as an appetite suppressant.[43] For every high-profile case of bioprospecting, there are many critics, with NGOs such as the Center for International Environmental Law taking on legal cases, and prominent scholars and activists such as Vandana Shiva contributing to the fight against widespread profiteering from Indigenous knowledge.[44]

In an attempt to protect Indigenous groups from the sharper end of bioprospecting, a regulatory framework known as the Nagoya Protocol was introduced in 2010.[45] As a result, it is now common for an agreement to be drawn up that ensures that some of the profits stemming from the commercial development of a biological resource be redistributed to the community in which it can be found.[46] In the case of olotón, the community who stands to benefit financially from an agreement with UC Davis and BioN2 is the village of Totontepec. While this is where the maize was supposedly "discovered" in the 1980s by Howard-Yana Shapiro, chief agricultural officer at Mars Inc., as noted, other varieties of indigenous corn have been seen to possess the ability to fix nitrogen elsewhere, including on Morales's own farm.[47] This raises questions as to who has the right to sell Indigenous knowledge and, when such sales do take place, how the resulting benefits are distributed fairly.[48] What is apparent, however, is the underlying dynamic that such agreements presuppose, encouraging two complementary acts of speculation: On the one hand, a company seeks out biological raw materials and gambles that the resulting product(s) will be worth its R&D expenditure and the expectations it has produced; on the other, a community, such as Totontepec, weighs up the pros and cons of signing such an agreement by speculating that a fraction of the profits will start to filter back into their community. In effect, both acts of speculation are underpinned by the international legal framework of intellectual property rights, which, as Noel Castree argues, "bioprospecting generalizes . . . as global 'common-sense.'"[49] A convention known as the UPOV '91 (UPOV stands for the International Union for the Protection of New Varieties of Plants, and '91 denotes the date on which it was redrafted) serves as a good example of this, and with Mexico's membership bundled into a trade deal with the United States, the intellectual property rights of plant breeders and ability of signatory governments to prosecute farmers who infringe these rights looks set to be bolstered significantly.[50]

In both her artwork and activist practice with groups, such as the Espacio Estatal en Defensa del Maíz Nativo de Oaxaca, who work at the intersection of Indigenous rights and environmental justice, Morales participates in what could be seen as an alternative infrastructure. This is one that runs counter to the "common sense" of commodification with an embedded practice that protects the cultural traditions that link plants to places. In *Aerial Root*, maize is framed as an "ancestral technology thousands of years in development." Yet in both the film and Morales's wider art practice, there is nevertheless an ambivalence to the visual techniques used to account for this socio-natural history. This is especially evident when the camera trains its sights on the olotón maize itself, and the bacteria-covered aerial roots that give the film its name. In a manner not dissimilar from Herregraven's use of commercial imagery discussed above, the high-definition close-ups of the gelatinous membrane (figure 2.5) that has earned the plant so much attention seem indistinguishable from the view of the bioprospector; and indeed, scientific publications about the plant contain remarkably similar images.[51] While the latter are accompanied by a breakdown of various chemical compounds measured in micrograms, there is a sense in which the visual language of Morales's film also participates in a gaze that visually isolates plant specimens and breaks them down into discrete biological functions, at the same time that it insists on the centrality of the milpa system and the cloud forests of Oaxaca as the cultural envelope in which such specimens belong.

What could be called the "specimen view" in this visual parsing of plant life serves as another example of an extractive gaze that resists any neat separation between "vertical" and "horizonal" orientations. The systematic collection, study, and classification of specimens is of course not restricted to plants, as histories of Enlightenment knowledge attest.[52] Indeed, O'Sullivan's approach to photography has itself been called "a kind of specimen collecting," insofar as it involved a process of observation and selection not dissimilar to the way the surveys' other specimen collections were formed.[53] Although common to many Enlightenment sciences,

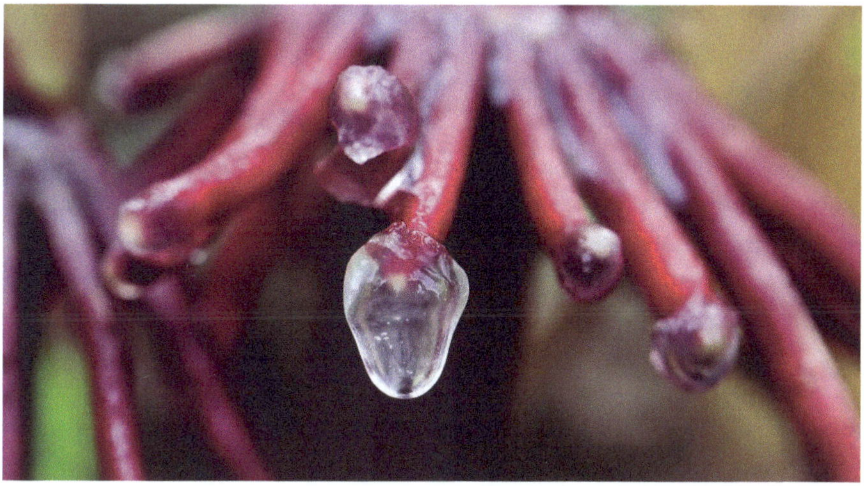

the isolation of individual species and their representation according to a set of conventions is central to the history of botany, particularly the operations of colonial botany. Here, voyages of discovery and scientific exploration often stripped plants of their cultural significance, "cleansing" them, as Daniela Bleichmar puts it, of their local, often spiritual importance.[54] This was done in the interests of uniformity: to create repositories of botanical information that were systematic and unvarying in their organization, and to facilitate comparison between specimens and easily retrievable information. Collections such as herbariums and xylotheques are exemplary in this respect, their function supported by such visual forms as the botanical illustration. As historians of natural science such as Kapil Raj show, these illustrations are governed by a set of strict conventions that require that "seeds be shown apart, whole and laterally dissected, that flowers be drawn separately, and that roots be depicted with

FIGURE 2.5

Edith Morales, still from *Aerial Root*, 2020.
Courtesy of the artist.

the plant."[55] Colonial botanical images are not produced with the primary goal of creating visual pleasure; rather, they convey critical information that can be operationalized in Western knowledge systems and mercantile economies of trade. Then as now, the specimen view is a visual form that facilitates speculation, as contemporary scientific reports on plants such as olotón maize attest. The conventions that govern the production of such images may be different today, but their basic function remains remarkably similar: starting with a plant that is seen as representative of its species, stripping it of its socio-ecological context, and further breaking this down into individual parts that fulfill specific functions.

In another artwork, Morales again trains her sights on the reproductive operations of maize, this time focusing on the grain itself. Her exhibition *Yoo Cua Nuniri / Autosuficiencia* at Parallel Oaxaca in 2019 featured, among other things, a work entitled *Altitud* (2019), in which several thousand corn seeds were mounted onto large white panels (figure 2.6).[56] The collection took the artist four years of painstaking research, sourcing varieties from remote regions of Oaxaca. Their arrangement for display, while seemingly random, is linked to the altitude to which the varieties have adapted. As Morales explained to me in an interview, corn tends to grow darker when cultivated 1,500 meters above sea level, rarely the familiar golden yellow of commercially available eating corn, developing in various shades of red, blue, and black.[57] The genetic variation is vast. Not only does an individual variety adapt to variables such as altitude and soil quality, but over time a variety will become its own landrace, with a distinct cycle of germination, growth, and reproduction. There were as many as twenty-seven distinct varieties in Morales's exhibition, each of which has deep cultural links to the communities in which it is found. The work is not simply a scientific demonstration of corn's adaptability but also a map of Oaxaca and the staple food on which many of its inhabitants depend.

Yet the significance of maize does not stop at the seed, or even at the plant. It is enfolded within a complex web of socio-ecological practices that have developed with and through its cultivation, as is already apparent in the milpa system. On a superficial reading of *Altitud*, the artist would seem to be stripping maize of its cultural context and perpetuating

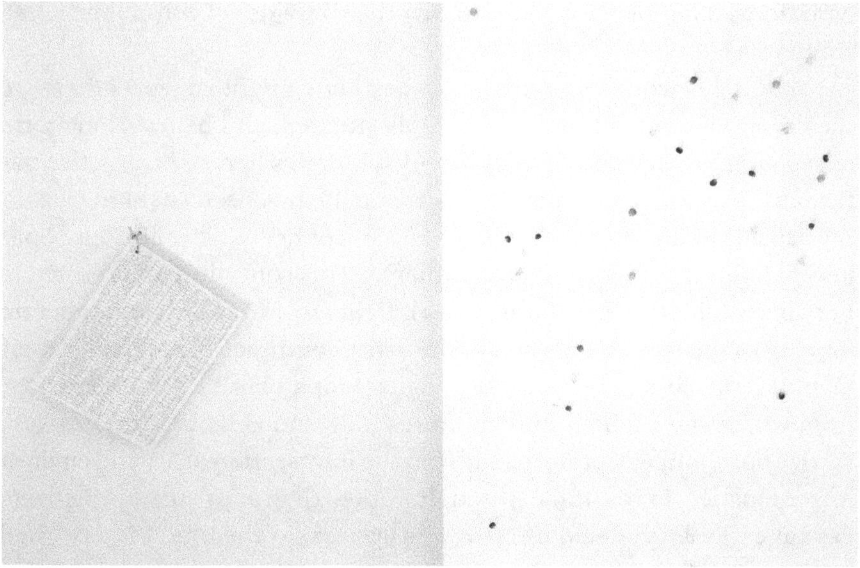

the kind of specimen view that permeates bioprospecting. A plant with immense cultural, historical, and ecological significance in Mexico is, at first glance, reduced to a collection of individual specimens, tabulated as data, and represented with the cold indifference of a commercial seed bank. Indeed, it is interesting to note that Morales works as an accountant in her day job, which might explain the aesthetics of administration present in many of her works.[58] The constellation of colored kernels, as attractive as they are, gives away very little sense of their significance, how they might grow in a field, or what they taste like in the mouth. When viewed in the context of Morales's work as a whole however, this detachment from the sensory particulars of the maize begins to feel somewhat

FIGURE 2.6

Edith Morales, *Altitud*, 2019. Courtesy of the artist.

parodic, as if deliberately exaggerating the strategies of commercial seed companies.

While this work's administrative aesthetic might suggest otherwise, the knowledge that informs it is deeply situated, and belies a long-term engagement with a community in which Morales herself is an active participant. As a Mixtec woman whose entire life has been spent in Oaxaca, the artist told me she measures the success of her work against the opinions of fellow activists and local farmers.[59] This commitment is rooted in her family history, with Morales's own father working as a farmer before moving to the city—a legacy that the artist continues through an experimental farm called the Centro Ecológico Milpa Urbana. This is the base from which she catalogs and cultivates indigenous maize varieties, and carries out a range of other activities at the intersection of art and environmental justice, for example, through work with groups such the Espacio Estatal en Defensa del Maíz Nativo de Oaxaca. At the time I interviewed her, Morales and many others were busy preparing for Mexico's accession to UPOV '91, which they expected to strengthen the hand of commercial plant breeders to claim varieties as their intellectual property. If the visual language of Morales's work sometimes echoes the detachment of such administrative and legal instruments, this serves to place these instruments in dialogue with the alternative knowledge infrastructures that the artist uses elsewhere in her work. Systems such as milpa existed long before the concept of owning the reproductive properties of a plant, the artist reminds us, and by understanding the extractive view that underpins this property relation, it might be possible to dislodge its claims to being common sense.

VISUAL CULTURE AT THE THRESHOLD

As theories of extraction are deployed with increasing frequency in the environmental arts and humanities, discussions of its accompanying visual culture have generated valuable analyses of the synoptic, rationalist, and reductive qualities of the aerial view. A study of "visual prospecting" extends these discussions, both by shifting the emphasis onto future resource extraction and by pluralizing the modes of visual culture prospecting can be seen to employ. While the three main examples discussed in this chapter are drawn from recent artistic practices, their differing inflections on the extractive view are already discernible in photographs produced over 150 years ago. The role of O'Sullivan's work in helping to frame the Western frontier as a territory of untapped potential is an early example of the conjuring act needed to attract speculative investment, as can be seen in the contemporary visual culture surrounding lithium extraction. Likewise, the "specimen view" discussed in the last section, which isolates items of economic interest and strips them of their socio-ecological relations, is also evident in the selection and presentation of O'Sullivan's survey photographs. Still more entanglements of extraction and visual culture exist in fields as diverse as deep-sea mining and space exploration, and it is a pressing task to understand the subtleties of how these come to structure forms of environmental exploitation that are deepening against the backdrop of climate change.[60]

While O'Sullivan's survey photographs compress many of the qualities of extractive vision discussed in this chapter, it is important to stress that they also differ in one crucial respect: O'Sullivan contributed to a visual culture *of* prospecting, while the other artists produce visual materials *about* prospecting. In other words, while the subject matter and visual strategies of the contemporary examples vary significantly, they represent artistic meditations on existing visual cultures of extraction—a "view on the view"—rather than images that can be operationalized to further prospecting's aims. By documenting the layered colonial histories of mining in the DRC or scrutinizing the bioprospecting activities of large food corporations against the backdrop of a changing climate, these artworks encourage reflection on the futures of resource extraction and how visual culture plays an active and potentially critical role within its production of knowledge and expectations. In the process, prospecting comes to be isolated as a speculative infrastructure with its own configuration of images, actors, and processes that contribute toward the unfolding story of environmental change.

Whether vertical, horizontal, or neither, visual prospecting is not a passive activity that gazes on an existing world of resources awaiting exploitation. The "extractive view" tied up in the search for biological and mineral riches is made up of a heterogeneous mix of tools and techniques that actively reconfigure the visual field. Artistic visual culture *about* prospecting is no less active, but instead of facilitating extraction, it often serves to mirror and subvert its visual language. In this way, the ever-shifting frontiers of extraction are spaces in which visual forms become malleable and open to reuse, where viewpoints or perspectives are not indentured to any one epistemological stance but rather bound up in processes whereby the threshold of what is knowable, seeable, and extractable are still being formed.

3

ENGINEERING

> We hope eventually to discern the great clanking wheels of the machinery in which man is such a small component. And perhaps in the end we can predict the times and places in which man's services will continue to be used or guess whether and when other parts will replace him in the drama of the earth.
>
> —Howard T. Odum, *Environment, Power, and Society*[1]

INTRODUCTION

What does it mean to engineer environmental futures into existence? How might engineering be separated from the avalanche of unintended consequences that capitalism, colonialism, and extractivism have unleashed on the climate? What knowledge infrastructures and visual cultures allow planetary systems to be conceived as engineerable objects in the first place? In order to answer these questions, this chapter opens with two contrasting examples, both of which demonstrate the effects of human intervention into marine ecosystems, albeit separated by a gulf of intention.

The first event took place in the spring of 1988, when a large and unexpected algae bloom occurred off the coast of Scandinavia. Dead fish were observed around the Skagerrak strait, a busy shipping lane lying between Denmark and Norway, in early May. By June, the algae had spread out of the strait and northwards along Norway's North Sea coast, traveling at an average speed of 20 to 25 kilometers per day and leaving a trail of decimated marine life in its wake. Around 500 tons of farmed fish were lost before commercial fisheries decided to evacuate their stocks, towing cages of salmon from the open sea into the brackish waters of the inner fjords.[2] Marine biologists quickly identified the algae responsible as *Chrysochromulina polylepis*, which was affecting the salt regulating activity of fish, creating a mucus that covered their gills.[3] It would take a little longer to identify the cause of the bloom, which was attributed to abnormally high concentrations of nutrients in the water, particularly nitrate, thought to have run off the land after a period of sustained wet weather.[4]

Several decades later, another species of algae bloomed in the Pacific Ocean; however, this time the bloom was created intentionally. Instead of killing fish, the anticipated effect of the Haida Salmon Restoration project was to boost dwindling supplies of pink and sockeye salmon off the west coast of Canada. Buoyed by the hypothesis that the fish population had increased several years previously when a plankton bloom was triggered by the deposition of volcanic ash from an eruption in Alaska, the private corporation responsible for the project dumped over a 100 metric tons of iron sulphate into the sea over a period of thirty days.[5] The resulting bloom, estimated at around 10,000 square kilometers in size, caused much controversy in the global press, and initial news reports contained allegations that the CEO and chief scientist on the project, American entrepreneur Russ George, had taken advantage of the Haida Nation Indigenous community in Old Massett village, who financed the project with a loan of 2.5 million Canadian dollars.[6] In fact, the picture was a lot more complex, with George signing on to the project in 2012 only after the Haida Salmon Restoration Corporation (HSRC) had been lent money by the Old Massett villagers.[7] As well as boosting salmon populations, an additional aim of the project was to sequester carbon through a technique known as ocean fertilization. The rationale for this was based on the theory that, since plankton absorbs CO_2 as it grows, when it dies and sinks to the seafloor, it inevitably takes some of this carbon with it—a theory that, despite its contested scientific status, did not stop George from attempting to monetize the process through the medium of carbon credits.[8]

The first plankton bloom in the North Sea was in part caused by human activity, with fertilizer run-off increasing the levels of nitrogen in the water. However, it was by no means planned, nor indeed desirable for any of the countries with coastlines bordering the affected region. By contrast, the second event was a deliberate act, with specific aims promoted to and by a defined group of stakeholders from the outset. While its actual effects are still unclear, in part due to doubts over the scientific rigor and transparency of the experiment, the Haida Salmon Restoration Corporation nevertheless sought to make a direct intervention into both the marine ecosystem of the Pacific Ocean and the carbon cycle as a whole.[9] It is this second intention that places the project in the politically fraught category of geoengineering. As one of its most prominent supporters puts it, "pollution is not engineering. Intent matters."[10]

An often-quoted definition of geoengineering describes it as the "deliberate and large-scale intervention in the Earth's climactic system with the aim of reducing global warming."[11] In recent years, narratives around geoengineering have proliferated in academia and mass media alike, making their way into mainstream media broadcasts and global news outlets, including the BBC, CBC, CNN, Al Jazeera, and many others.[12] For every promethean endorsement of tech-centered responses to climate change, there are just as many voices critical of the idea of assuming the position of "Earthmasters" able to make direct interventions into a climate system still not fully understood.[13] This chapter contributes to this critical literature, but it does so not by sifting through the various tools and technologies associated with geoengineering to evaluate their relative merits.[14] It begins, rather, by looking at the techno-scientific imaginary of the planet as a series of interconnected systems that can be programmed and modified and that allows us to speak of "engineering" the climate in the first place.

The first example is significant in another respect, because it demonstrates the importance of a certain mode of visuality associated with geoengineering. This devastating ecological event in the North Sea quickly snowballed into a media event, with satellite imagery and remote sensing data feeding into daily forecasts that were distributed to radio, newspapers, and television. The artist Peter Fend, as part of the organization Ocean Earth Development Corporation, had a front-row seat for the spectacle.[15] Registered as a commercial enterprise in the US, Ocean Earth had carved out a unique niche supplying satellite images to the media, finding notable success several years prior in the run up to the Falklands conflict off the coast of Argentina. For the algal bloom of 1988, Ocean Earth provided images to Dutch TV, mobilizing the power of the media and new image processing technologies to put pressure on government officials (figure 3.1).[16]

The use of satellite imagery is one example of what could be called a "macroscopic view" on the environment, which far exceeds what could be generated by individuals or groups gathering information through other means. At the core of this chapter is an exploration of this macroscopic view and its relationship to geoengineering, predictive rationality, and artworks that make its representational logic visible. While satellite imagery has played an important role in expanding the macroscopic view, its aerial or vertical orientation is not the defining characteristic here. As the word "macroscopic" implies, I argue that what is more significant is the scale and resolution at which it operates. Like the extractive view discussed in the previous chapter, the macroscopic view has an ambivalence that enables it to simultaneously be mobilized in the services of environmental monitoring and conservation and lay the groundwork for climate modification proposals on a planetary scale. It is informed by a series of abstractions that are essential to modern environmental science, insofar as they have facilitated the creation of successive "data images" that have converged into a global scientific consensus on climate change and the futures it entails.[17] The abstractions at the root of geoengineering have also, this chapter argues, contributed to a mode of speculation that has a tendency to close the gaps between representation and reality, translating models into policy prescriptions and sidelining the future-making capacity of more-than-humans in the process.

The artworks and design propositions discussed in this chapter provide points of orientation through this ambivalent terrain, facilitating visual literacy and providing situated perspectives from which to see at scale. To varying degrees, creative practices can visualize what happens when the macroscopic view is endowed with the power to make worlds, and how this power becomes embedded within socio-technical systems that give rise to a specific mode of relation to futures. At its core is an

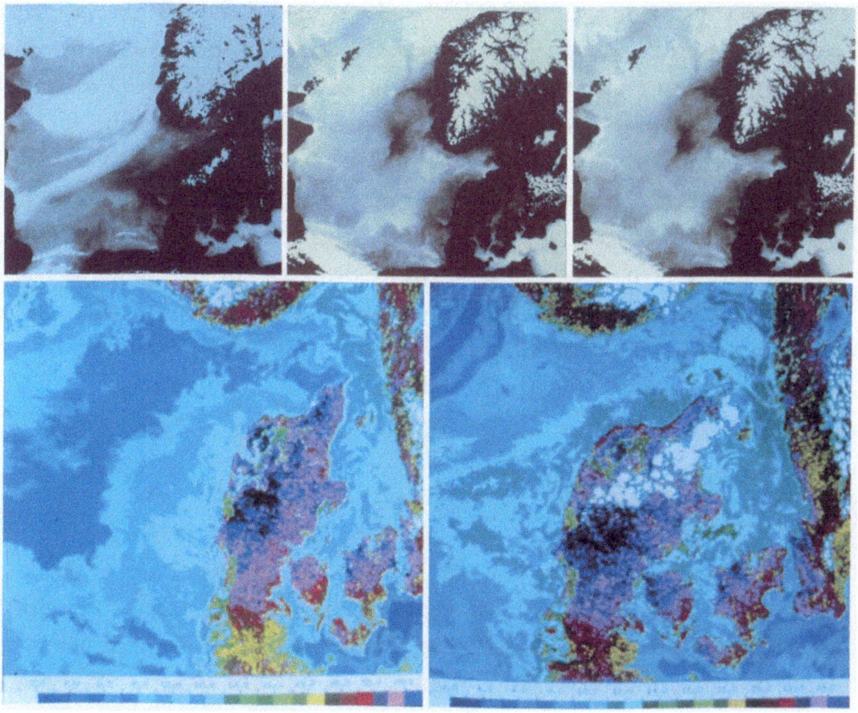

FIGURE 3.1

Ocean Earth Development Corporation, "Raw Imagery from AVRR of the North Sea, May 1988" (top three images) and "Processed imagery from AVHRR of the North Sea, 15–16 May 1988, showing a change in a 24-hour span, and explosion of the algae bloom from the small point where it began" (bottom two images). Images appear in Ocean Earth, *Ocean Earth: 1980 Bis Heute*, ed. Peter Weiber et al. (Neue Galerie am Landesmuseum Joanneum, 1994), 117.

assertion that to understand the visual culture of geoengineering, it is necessary to look past the spectacular sci-fi aesthetic of space mirrors and cloud seeding programs, and instead unpick the visual cultures that translate more-than-human ecologies into system diagrams and models that have significant consequences for how scientists, militaries, and policymakers imagine their ability to engineer planetary futures into being. Toward the end of the chapter, inspiration is taken from two designers who explore alternatives to this visual culture of the macroscopic view and, I argue, suggest ways in which areas particularly vulnerable to the effects of climate change, such as Mumbai, might serve as points of orientation for a more situated visual culture of preparedness.

A PLAN FOR THE PLANET

Ocean iron fertilization is one of many proposed responses to climate change that seek to treat the symptoms of greenhouse gas emissions rather than attempting to reduce their production at the source. The simplicity of the definition of geoengineering quoted above—the "deliberate and large-scale intervention in the Earth's climactic system with the aim of reducing global warming"—belies a diverse field of activity, ranging from relatively uncontroversial proposals, such as afforestation, to more experimental techniques, including "enhanced weathering" (mining, crushing, and spreading large quantities of minerals to accelerate the natural weathering process that draws CO_2 from the atmosphere) and "stratospheric aerosol injection," which aims to increase the reflectivity of the earth's atmosphere by spraying large quantities of sulfur or other aerosols into the air.

In the late 2000s, the field of geoengineering was deemed so muddled that a group of researchers gathered to try to impose some order on the chaos. Citing the problems of a field "bedevilled by much doubt and confusion," the director of the Royal Society, Martin Rees, framed the resulting report in language typical of one of the most prestigious institutions of science: as an attempt to "provide an authoritative and balanced assessment of the main geoengineering options" being discussed.[18] In the face of much epistemological confusion, one of the lasting effects of the report was to make a clear distinction between techniques of carbon dioxide removal (CDR) and those of solar radiation management (SRM). In basic terms, where SRM is about reducing the amount of sunlight that reaches the earth, for example, by reflecting more of what does pass through the ozone layer back into space, CDR aims to suck CO_2 from the air and store it somewhere it cannot re-enter circulation. Of the two categories, the Royal Society lent their cautious support for CDR, shying away from an endorsement of the comparatively riskier SRM technologies, such as stratospheric aerosols, which were nevertheless described as the closest thing to "an ideal method," in terms of the expected balance between effectiveness and affordability.[19] The report appeared in 2009, and in the almost two decades following its publication, some of the caution around endorsing both SRM and CDR has melted away, with assumptions around the future effectiveness of so called "negative emissions technologies" now routinely integrated in climate models, and the IPCC providing a summary for policy makers of geoengineering options in one of its reports.[20] In terms of field trials, a thirty-million-pound research project into CDR technologies in the UK ran from 2021 to 2025, led by an environmental economist and former Shell employee.[21] A smaller project into SRM techniques began at Harvard in 2019, only to be abandoned in 2024 following public criticism, not least from the Saami Council, which represents Indigenous communities of northwest Russia, northern Scandinavia, and Finland, where one of the trials was scheduled to take place in 2021.[22] Experiments in "marine cloud brightening" have taken place in both California and Australia since 2020, the latter as an attempt to reduce coral bleaching.[23] Taken together, these publications, trials, and the institutional groupings that produce them constitute an emerging knowledge infrastructure for geoengineering.

CDR technologies have also filtered down into the art world, with Daan Roosegaarde's *Smog Free Tower* (figure 3.2) serving as an eye-catching advertisement for the burgeoning cleantech sector in a number of international cities, such as Rotterdam, Beijing, and Krakow. Using a patented ionization technique, the artist's bulky obelisks are designed to suck up pollution in urban locations such as city parks in order to create "clean air bubbles." No less entrepreneurial in their ambitions, what first differentiated Roosegaarde's towers from other cleantech innovations is

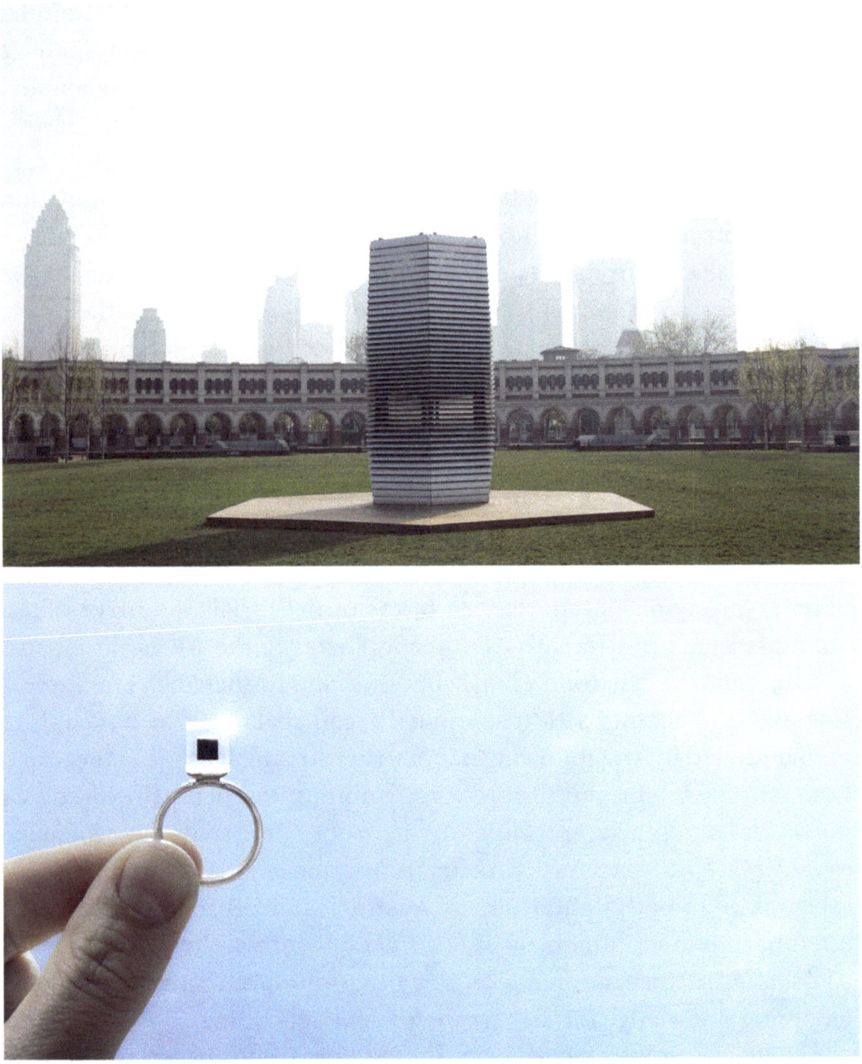

FIGURE 3.2

Daan Roosegaarde, *Smog Free Tower*, installed in Tianjin, and *Smog Free Ring* (both 2017). Courtesy of Studio Roosegaarde.

their ability to capture smog particles and then put them under enough pressure to transform them into items of jewelry. The resulting objects not only act as miniature carbon storage solutions but also convenient rewards for the studio's crowdfunding campaign, which managed to garner financial contributions exceeding 100,000 euros shortly after it was launched in 2015. Over a decade later, others are pursuing similar plans, with the company Aether specializing in "chemical vapor deposition diamonds" that are now produced for a larger customer base.

The eye-catching alchemy of *Smog Free Tower* is one of many localized interventions that have helped normalize the idea that the effects of climate change can be treated downstream, as a problem of waste management. Operating on a rather different scale, the Nobel Prize–winning atmospheric chemist Paul Crutzen triggered a surge in publications on geoengineering in 2006 when he suggested that technologies such as stratospheric sulfur injections could provide an "escape route" out of climate change and cannot be ignored.[24] Drawing on ethnographic research, Stilgoe shows that this influential publication, together with the Royal Society report several years later, are considered by many researchers in the field to be key moments that transformed geoengineering from a collection of "potent, troublesome technologies, segregated from mainstream science by a fragile taboo" to legitimate objects of research.[25] This framing of the issue typically asserts that in the early 2000s, geoengineering technologies were so new, so experimental, and so fraught with moral hazards that even discussing them as options undermined attempts at reducing emissions at source—until, that is, the Crutzen and Royal Society publications let the cat out of the bag.

As Stilgoe argues, the narrative of a taboo on geoengineering research being gradually lifted is problematic in several respects. For one, the term "geoengineering" does not describe a storehouse of potent technologies waiting for social permission to be deployed. Many of the ideas under discussion are theoretical at best, with little empirical evidence to demonstrate claims of their effectiveness at scale. Some proposals, such as the SRM strategy of installing mirrors in space to reflect sunlight, are considered so impractical as to attract ridicule, being the products of a techno-positivism that blurs the boundaries between Hollywood spectacle and the kind of spectacle needed to attract investment. Secondly, for all the comparisons to other fields of research whose novelty marked watershed moments in the history of technoscience—synthetic biology and nanotechnology, for example—many of the ideas that fall into the category of geoengineering are neither as new nor as groundbreaking as many of their proponents suggest.

In his history of weather and climate control, *Fixing the Sky*, James Fleming traces a line of continuity between Cold War attempts to control the weather and more contemporary geoengineering proposals. Techniques such as stratospheric aerosol injections are in fact much older than many might assume, with Soviet proposals "to create a dust screen to reduce solar radiation" dating back to the early 1960s, and scientists on the other side of the Iron Curtain ruminating on the possibility of changing the albedo effect by deploying "microscopic layers of colored matter" onto large bodies of ice.[26] These ideas also entered the public realm through popular science fiction of the 1960s and '70s, with Ben Bova's *The Weathermakers* (1967) exploring cloud seeding and weather modification with laser technology, and J. G. Ballard's "Cloud-Sculptors of Coral D" (1967) providing a vivid image of the technology being put to aesthetic use.[27] Often these imaginaries, whether science fiction or fact, were not explicitly linked to an awareness of climate change—a topic that had nevertheless made its way into policy discussions in the 1960s, albeit usually within reports on broader or adjacent subjects.[28] While the systematic measurement of atmospheric CO_2 by figures such as Charles David Keeling was already underway in the late 1950s, his findings would not converge with those of others into a scientific consensus until over a decade later, long after the first proposals for weather modification.[29]

However, when it came to modifying the weather for military gain, there was enough momentum behind the idea to trigger far-reaching experiments in the real world. "Operation Popeye" was one such experiment. Executed during the Vietnam War, the aim was to extend the monsoon season through repeated cloud seeding carried out over several years, from 1967 to 1972.[30] The resemblance of many geoengineering initiatives to these Cold War–era experiments is not incidental for Fleming, who sees the intellectual culture cultivated during this period as having a lasting influence on geoengineering research.[31] Clive Hamilton reinforces this by showing how some of the same institutions were involved in both areas of research, with the continued militarization of the geoengineering agenda discernible in reports published by organizations such as RAND, produced both in the 1960s and in the past decade.[32]

Despite the numerous historical precedents, both actual and imagined, many current day geoengineering proposals combine hype and hubris to create an aura of novelty. As Kathryn Yusoff writes, "What geoengineering gains in loosening its historical ties to weather modification projects of the 1950s is an ability to claim its precedence, as a 'new' solution to climate change."[33] The appearance of novelty in technoscience is a key ingredient in producing expectations and attracting the necessary speculative capital to carry projects forward, as discussed in the context of prospecting in the previous chapter. One of the highest-profile advocates of geoengineering in the US, Ken Caldeira, holds a range of patents on geoengineering technologies ranging from "Mitigating global warming by OTEC-induced ocean upwelling" (2015) to "Extracting and sequestering of carbon dioxide from a gas stream" (2009), and an Australian company has even claimed to own the fish produced by their trademarked Ocean Nourishment™ technology.[34] The amount of proprietary knowledge in current geoengineering research has led the online video channel "The Inhabitants" (an initiative launched by the artists Pedro Neves Marques and Mariana Silva in 2015 as a vehicle for distributing experimental short-form documentary reportage on topics including deep-sea mining and oil exploration) to speak of a "private-led patent rush" that secures the interests of venture capital pouring into the field.[35] A large proportion of this money comes from companies who stand to gain if the agenda shifts toward negative emission technologies and SRM techniques and away from mitigation measures that could severely impact their profits.

MACROSCOPES AND MICROSCOPES

What worldview enables one to speak of engineering something as immense as the climate? And how do visual cultures that scale play a part in constructing the biosphere as a "vast machine" that can be subject to technologies of rationalist control?[36]

It is not insignificant that the first article to use the word "geoengineering" in something like its current day sense was authored by an early adopter of predictive modeling, the Italian physicist Cesare Marchetti. During his time at the International Institute for Applied Systems Analysis (IIASA), an organization set up in the 1970s to explore the application of cybernetics and modeling techniques to matters of global concern, Marchetti set out a proposal for CO_2 to be captured and injected "into the deep seas taking advantage of natural thermohaline circulations."[37] What is of interest here is not simply that geoengineering proposals have an unexpectedly long history, but also that the framing of the world as an object amenable to systemic analysis has been stubbornly persistent since the Cold War. It was during this period that a vision of the planet as a series of interconnected feedback loops emerged from the epistemic culture of cybernetics, systems analysis, and operations research.[38] Such rationality provided a means of conceptualizing the biosphere, in many ways laying the groundwork for a new form of predictive environmental expertise and a sense of the planet as a fragile system of interdependencies.

Systems rationality would also provide some of the key ingredients for Gaia theory, first proposed by James Lovelock in collaboration with Dian Hitchcock and later developed with extensive input from Lynn Margulis.[39] According to this theory, the planet should be understood as a self-regulating system whose interacting components work together to perpetuate conditions suitable for life. Freely drawing on cybernetic metaphors, Lovelock and Margulis spoke of atmospheric processes in terms of "homeostasis" and "feedback loops" that contributed to the maintenance of life on earth. This vocabulary was sometimes complemented by Lovelock with biological metaphors, notably describing the earth as "a superorganism"—a term that Margulis later disputed, preferring the more scientific language of ecosystems.[40] As others have noted, the roots of Gaia theory have a thoroughly "top-down" origin story, grounded in Lovelock's time working for NASA, where his research focused on finding atmospheric disequilibrium on other planets, principally Mars, that would indicate life.[41] While there is another, more "bottom-up" origin story for the hypothesis that follows the implications of microbial life, this top-down perspective is nevertheless important in placing Gaia theory in the intellectual culture of the time, a culture that also gave rise to

Earth System Science.[42] As Eric Conway notes, at least in their early formulation, "both Gaia and Earth System Science were based in a systems engineering view of the world . . . a profession that specialized in (nonliving) feedback control systems."[43]

The common thread uniting present-day geoengineering research, early proposals by the likes of Marachetti, the "top-down" story of Gaia theory, and Cold War attempts at weather modification is a "macroscopic view" of the planet and the living matter that inhabits it. This perspective, overlapping and yet distinct from the aerial view discussed in the last chapter, is inextricably entangled with the history of ecology as a discipline. This becomes abundantly clear in the work of Howard T. Odum, who together with his brother Eugene would help establish the discipline of ecology in its modern form in the mid-twentieth century, laying the foundations for Gaia and Earth System Science in the 1960s. The Odums envisaged the ecologist's role as one aided by an imaginary "macroscope" that would "eliminate the details" of the individual organisms and their complex interdependencies.[44] In the original edition of his expansive tract on ecological thinking, *Environment, Power, and Society*, the younger of the two Odums wrote in broad terms about the need to gain perspective by zooming out, rather than focusing in:

> Whereas two centuries of scientific progress were derived from the microscopic work, we find the contemporary world beginning to look through a macroscope of systems science and acquiring ways to discern the broad features and mechanisms of a system of parts. Bit by bit the machinery of the macroscope is evolving in various sciences and in the philosophic attitudes of students. The daily maps of worldwide weather, the information received from the high-flying satellites, the macroeconomic statistical summaries of nations and the world, the combined efforts of international geophysical collaborations, and the radioactive studies of cycling chemicals in the great oceans all stimulate the new view.[45]

For the Odums, this kind of big-picture thinking enabled the vital information regarding ecosystems to be looked at in isolation from the real-world detail deemed unnecessary. The path to abstraction was one that passed through the filter of quantification, and the Odums sought wherever they could to translate ecosystem dynamics into units of energy, which flowed along "the invisible wires of nature."[46] In many cases they complemented this work with research of a more biochemical nature, following the metabolic pathways through ecosystems by introducing radioactive tracers and seeing how they circulated.[47] As well as drawing on atomic research, the Odums also famously produced circuit diagrams of their findings using a visual language that drew explicitly on cybernetics and electrical engineering. Here, most if not all exchanges were reduced to transfers of energy that could be quantified in units such as kilocalories (figure 3.3). But the brothers were not only interested in producing macroscopic representations of the world they observed. Looking through a "detail eliminator" was merely the first necessary step along a path toward managing the environment with concrete actions, as a cartoon illustration produced by Howard indicates (figure 3.4). In his words, this was a path from "explanation, [to] prediction and control."[48] If ecosystems could be correctly translated into flows of energy, the component parts could be readily treated as "an immense bin of parts available to the ecological engineer," who could then reconfigure them with specific objectives in mind.[49] The position of the ecological engineer is significant here,

FIGURE 3.3

Energy circuit diagrams for tribal cattle system in Uganda, appearing in Howard T. Odum, "Energetics of World Food Production," in *Report of the Panel on World Food Supply*, vol. 3 of *The World Food Problem: A Report of the President's Science Advisory Committee* (White House, 1967), 72.

FIGURE 3.4

A cartoon illustrating the "macroscopic view," first appearing in 1971, and reprinted in Howard T. Odum, *Environment, Power, and Society for the Twenty-First Century: The Hierarchy of Energy* (Columbia University Press, 2007), 10. Reprinted with permission of Columbia University Press.

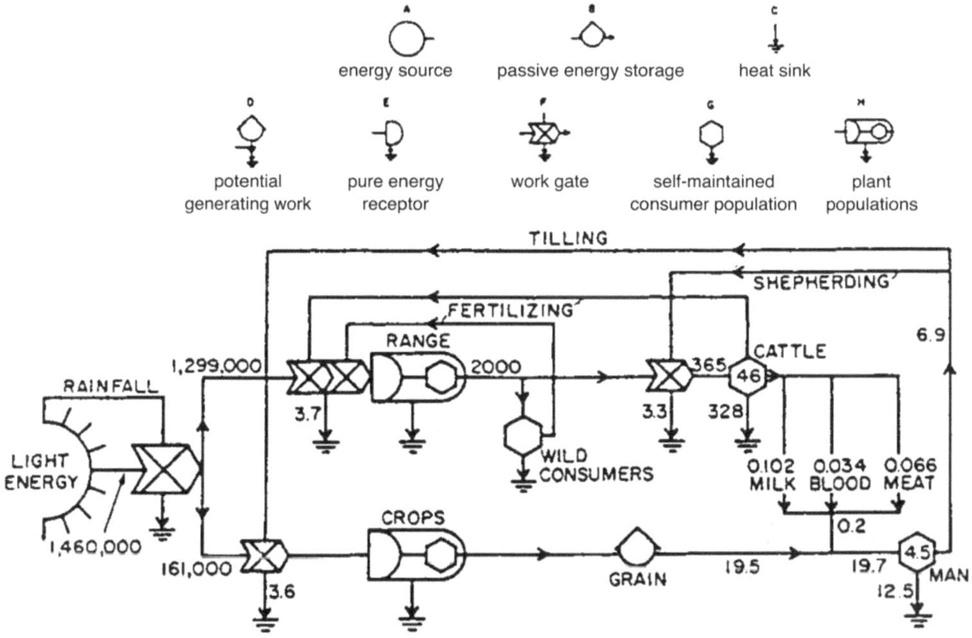
A
B
C
energy source
passive energy storage
heat sink
D
E
F
G
H
potential generating work
pure energy receptor
work gate
self-maintained consumer population
plant populations
TILLING
SHEPHERDING
FERTILIZING
RANGE
CATTLE
6.9
1,299,000
2000
365
46
RAINFALL
3.7
3.3
328
LIGHT ENERGY
WILD CONSUMERS
0.102 MILK
0.034 BLOOD
0.066 MEAT
CROPS
0.2
1,460,000
161,000
3.6
GRAIN
19.5
19.7
4.5
MAN
12.5

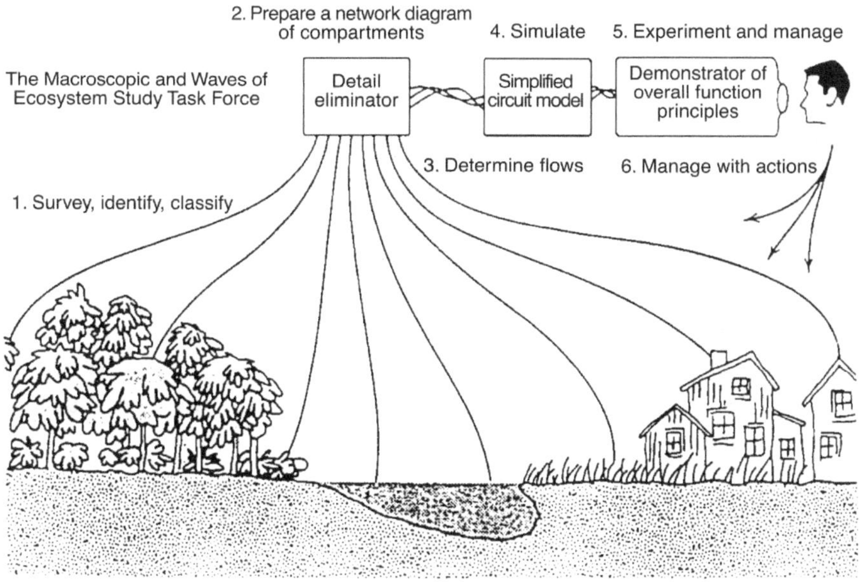
2. Prepare a network diagram of compartments
4. Simulate
5. Experiment and manage
The Macroscopic and Waves of Ecosystem Study Task Force
Detail eliminator
Simplified circuit model
Demonstrator of overall function principles
3. Determine flows
6. Manage with actions
1. Survey, identify, classify

because it implies a position above or outside the world, emblematic of the "godtrick of seeing everything from nowhere" that Haraway famously contested by underscoring the importance of situated knowledge.[50]

It is doubtful that the Odums would have lent their support to such risky strategies as injecting the stratosphere with large quantities of sulfur, but their influential systems ecology of feedback loops and energy transfers nevertheless paved the way for the kind of thinking that treats the biosphere as an object that geoengineers can modify with predictable outcomes at a planetary scale. Making a similar point, Kathryn Yusoff identifies the ontological and epistemological foundations of geoengineering in the construction of an earth "in which atmospheres and earth systems, and even the sun, are imagined as 'engines' (albeit organic) open to forms of making."[51] Seeking to distinguish this conceptualization of the earth from other macroscopic perspectives, such as those inspired by the blue marble images mentioned in the previous chapter, Yusoff emphasizes that this way of thinking about the planet is not one of something found—an artifact—nor indeed of a technology with fixed inputs and outputs, but of a "bio-geophysical machine," in which flows of energy and information are treated as amenable to intervention.

Such conceptions were not restricted to thinkers in America and Western Europe in the twentieth century. Several decades before the Odums stressed the importance of a macroscopic view from their base in American academia, the Russian geochemist Vladimir Vernadsky was developing an idea of the "noosphere" (first coined by the Jesuit philosopher Teilhard de Chardin) that spoke directly to ideas of planetary management, insofar as it denotes the emergence of a collective human consciousness.[52] In abbreviated form, Vernadsky's concept of the noosphere identifies a stage of planetary development in which human cognition is able to transform the biosphere, in the same way that life on earth shaped its geological conditions by contributing oxygen, nitrogen, and carbon dioxide to the atmosphere. In the 1970s and '80s, the concept was picked up and reworked by the mathematician Nikita Moiseyev, an early adopter of systems analysis and global modeling, who thought it could have significance in the context of environmental governance.[53] Like Marchetti, Lovelock and Margulis, and the Odum brothers, Moiseyev was also adept at conceptualizing environments as systems of inputs, outputs, and feedback loops, and much of his later work uses the concept of the noosphere to consider how such a perspective might be translated into governing planetary processes. According to Eglė Rindzevičiūtė, what marks Moiseyev's contribution as distinctive is his nuanced appraisal of how scientific predictions based on such ideas are fed into the machinery of government. For Moiseyev, the knowledge produced by systems analysis and global models is simply too crude to produce anything close to the kind of certainty required for policy prescriptions at a global scale, and therefore, the only way of putting this knowledge to use is by translating it into "negative guidance."[54] Faced with knowledge of the planet that is uncertain and partial, the answer was not to provide policy prescriptions, as some supporters of geoengineering would advocate, but rather to impose limits and prohibitions on what is considered viable.[55] While it is not possible to do justice to his nuanced approach here, Moiseyev's work nevertheless represents an important variation on ideas around planetary environmental governance and the macroscopic perspective from which geoengineering emerges.

While Moiseyev was careful to point out the limits of predictions produced at a planetary scale, others show less reluctance to suggest far-reaching interventions. As well as Marchetti, it is notable that, late in his career, Lovelock also proposed ocean fertilization as a "way to help the earth cure itself," retreating from the idea in the years that followed.[56] The macroscopic view upon which such proposals depend has also been shaped by countless artists, including in the example that opened this chapter by Peter Fend, whose planetary ambitions are already evident in the name "Ocean Earth Development Corporation." The niche that Ocean Earth carved out for itself procuring, analyzing, and selling satellite-based imagery was as much about cultivating a new way of seeing as about producing tangible artworks. Born out of a frustration with what he describes as "Renaissance-based procedures for trying to address problems one by one, down perspectival alleys to vanishing points," Fend and his collaborators embraced the emerging possibilities offered by satellite images.[57] Monitoring the earth with these tools was not a purely descriptive or analytical exercise; for Ocean Earth, the end goal was nothing short of the "planning and maintenance of the planet"—a goal that nevertheless remained far beyond the reach of the group in any meaningful way.[58]

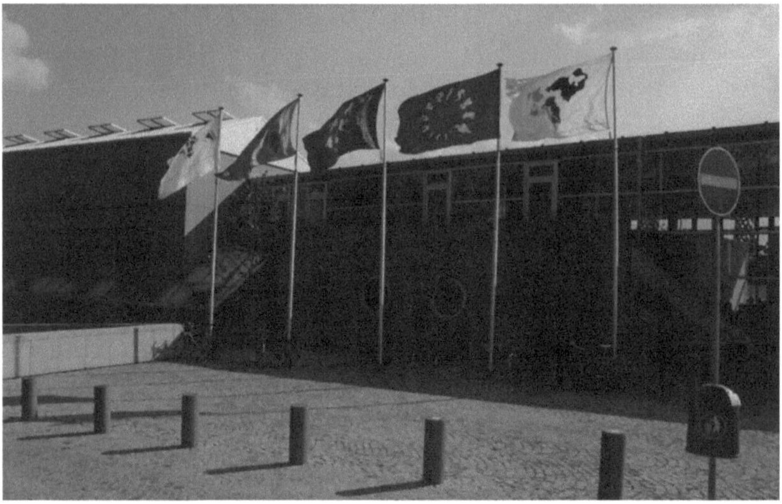

Indicative of the scale of Ocean Earth's bold ambitions, the group has proposed on several occasions that much of the earth should be reorganized into political units dictated by the geography of watershed basins, most notably in 1992 at Documenta IX, where Fend and his collaborators replaced the usual display of national flags with their own creations featuring bold graphic representations of imaginary political entities (figure 3.5). Here an ecocentric political stance is proposed at a macro scale in a way that sidelines any consideration for the nuances of national, ethnic, and religious identity in any given region, revealing some obvious blind spots to such a proposal.[59] A similar idea features in one of Fend's first solo exhibitions, *Wealth of Basins, Death of Nations* in 1979, its title a wry parody of Adam Smith's magnum opus, which is contrasted to an alternative history of economic thought dating back to the physiocrats of eighteenth-century France.[60] Fend explains the contemporary relevance

FIGURE 3.5

Ocean Earth Development Corporation, *Flags for Documenta*, 1992. Installation View at Documenta IX.

of physiocracy in an era when income tax is easily avoided by those with enough resources to exploit loopholes. Instead, he has proposed a form of taxation based on the ownership and use of land, measured in the pixels of satellite imagery. Forms of land value tax already operate in many parts of the world, but the ambition of Fend's proposal lies in its geo-planetary scale and dependence on the aesthetic properties of "overview field studies" and "abstract conceptions" rather than a concern for localized sociopolitical conditions on the ground.[61] Fend is confident of the impartiality of satellite data in this respect, which he sees as having "an evidentiary facility" and "universality," insofar as it cannot be easily tampered with.[62]

The macroscopic view underpins virtually all of Ocean Earth's projects, both those originating in the art world and their work supplying satellite images to media organizations. Given this long-held preoccupation, it is perhaps surprising that Fend has been critical of initiatives such as Google Earth, which one might otherwise expect him to celebrate. Fend sees Google Earth, instead of having useful applications, such as using satellite imagery to track dynamic changes in the distribution and functioning of socioecological systems, as more concerned with the narcissistic satisfaction of looking at ourselves.[63] In many ways, the work of Ocean Earth can be seen as a corrective to this narcissism; by proposing large-scale political reorganization based on river systems and saltwater basins, populations are realigned with the workings of nature understood as a giant bio-geophysical machine, a proposal with anti-humanist dimensions that could tip into ecofascism if left unqualified.[64]

Ocean Earth's ecocentric stance does not prevent much of their work from resembling current-day geoengineering initiatives either, which can be seen as symbols of humanity's mastery over the environment.[65] One idea that resurfaces in various Ocean Earth projects is for the creation of a new form of "mariculture" that combines farming of species of algae for the purposes of producing biofuel with the idea of capturing wild populations of fish using a semi-permeable system of nets. Fend's first concept sketch for the idea of an "oceanic micro-algae rig" appeared in 1979. Like Marchetti's early proposals for deep-sea carbon storage, if built, the rigs would take advantage of the natural flows of the ocean, by being placed where nutrient cycling and fish stocks are highest.[66] In an interview, Fend tells me that New York City receives "one of the richest bodies of ocean water in the world, coming out of the Arctic," and that it is "simply a matter of taking advantage of it."[67] The idea was explored in the Adriatic Basin in 1989, when Ocean Earth made a typically ambitious proposal that combined "panoptic productivity monitoring" of the sea using satellites and remote sensing devices with a form of marine algae cultivation that would produce enough methane gas to significantly weaken the dependency of neighboring industrial markets on fossil fuels.[68] A few years later, a more detailed set of the proposals surfaced in the exhibition *France: Strategie Globale* at FRAC Poitou-Charentes, Angoulême, in 1993, which involved scale models of the rig developed together with the naval architect Mark Lombard (figure 3.6). Fend's frustration that such models have not been translated into reality is palpable, citing numerous instances in which the art world has constrained him to producing representations rather than fully functioning objects.[69]

Similar proposals involving the cultivation of algae as a biofuel feedstock remained unrealized for different reasons, beginning in 1972 with the "Ocean Food and Energy Farm Project," funded in part by the US Navy (figure 3.7).[70] In her book *After Geoengineering*, Holly Jean Buck explores the development of such projects over the past fifty years, looking at how the idea of cultivating and harvesting seaweed originated in response to the rapidly increasing price of fuels in the 1970s and '80s.

FIGURE 3.6

Peter Fend/Ocean Earth, seaweed cultivation rig, 1993. Installation view at FRAC Poitou-Charentes, Angoulême, 1993.

FIGURE 3.7

Howard A. Wilcox, "Conceptual design. 400-ha ocean food and energy farm unit" (1973/1982), in Howard A. Wilcox, "The Ocean as a Supplier of Food and Energy," *Experientia* 38 (1982): 33. In this article Wilcox writes that the concept of the food and energy farm unit had been under consideration since 1972.

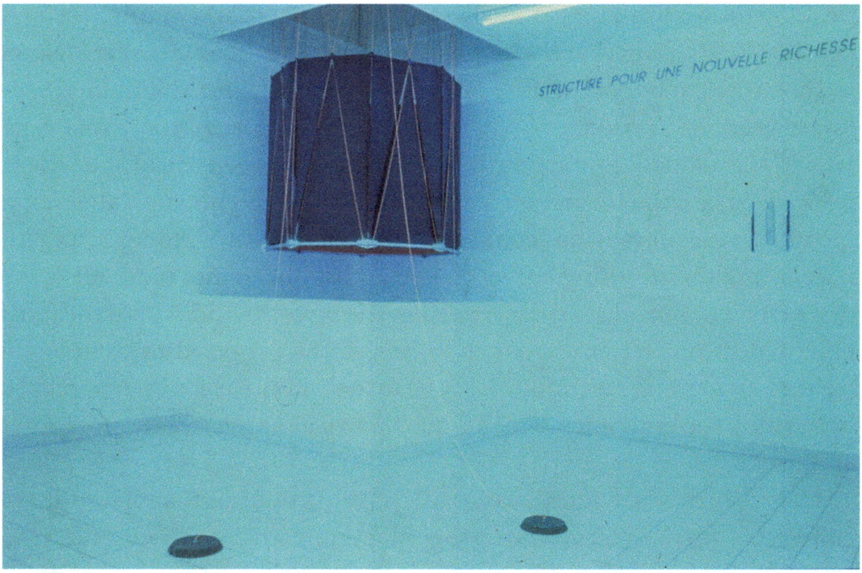
STRUCTURE POUR UNE NOUVELLE RICHESSE

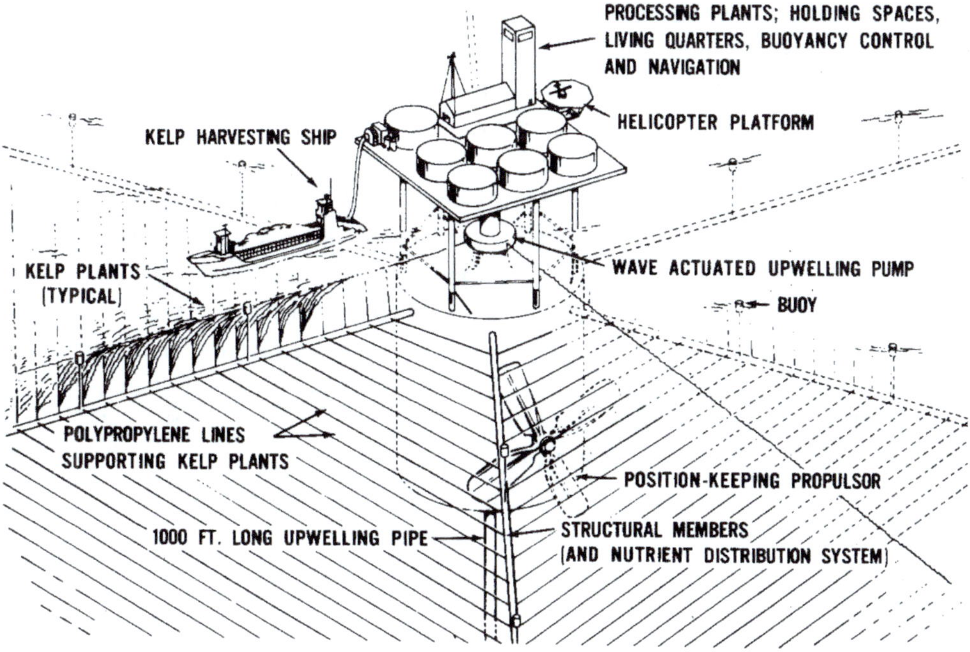
PROCESSING PLANTS; HOLDING SPACES, LIVING QUARTERS, BUOYANCY CONTROL AND NAVIGATION
HELICOPTER PLATFORM
KELP HARVESTING SHIP
KELP PLANTS (TYPICAL)
WAVE ACTUATED UPWELLING PUMP
BUOY
POLYPROPYLENE LINES SUPPORTING KELP PLANTS
POSITION-KEEPING PROPULSOR
1000 FT. LONG UPWELLING PIPE
STRUCTURAL MEMBERS (AND NUTRIENT DISTRIBUTION SYSTEM)

Deemed unfeasible at the time due to the requisite costs of complex engineering and the fact that the kelp farms would have to each cover areas of 100,000 acres to be competitive (a size now not unrealistic with increases in seaweed cultivation globally, particularly in China), the plans were largely abandoned.[71] Charting something of a renaissance of such ideas, Buck discusses the impact on feasibility of recent technological developments. Drone submarines could tow smaller kelp farms to nutrient rich waters, and the resulting harvest could be converted into biocrude using more efficient processes than existed in the 1970s. The idea with the most potential to impact emissions, however, is the proposal to use carbon capture and storage techniques (CCS) to lock away the emissions caused by burning the biocrude—a technique known as seaweed carbon capture and sink.[72]

It is only this last addition to mariculture proposals that places ocean cultivation in the category of geoengineering. Ocean afforestation, as some of its advocates call it, may have a positive effect on food security, or even improve marine ecosystems, but without the addition of CCS, it will have little effect on climate change. Nevertheless, what all these ideas share is a mutual investment in a macroscopic view of the earth as a dynamic bio-geophysical system, whereby the focus is on policy prescriptions centered around technological interventions rather than meaningful changes to economic structures of growth and accumulation.[73] Even Ocean Earth's proposal for political units based on watershed basins leaves these fundamental structures untouched. Fend's decision to establish Ocean Earth as a for-profit company was not an art-world gesture to attract attention. It is underpinned by a strong belief in the power of private enterprise and motivating force of competition, unimpeded by the inertia of central governments. Explaining his plan to take out patents on a number of his ideas, he claims that Ocean Earth has the potential to be "a multi-billion dollar company," albeit perhaps not in his lifetime.[74] This confidence could be seen as an artifact of the macroscopic view, which constructs as its object a model world untroubled by the details of implementation. Here, environmental futures are approached in ways that collapse the gap between reality and representation, neglecting the unexpected outcomes they would inevitably encounter in the world seen at full resolution.

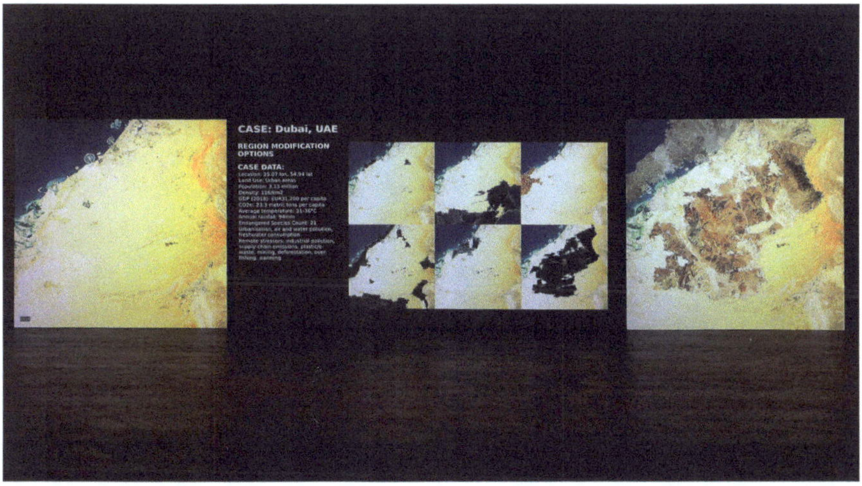

CRITICAL ENGINEERING

In a large room in Vienna, a 144-CPU super-computer is at work. Sitting on a slender plinth with its internal components exposed, the computer proposes solutions to some of the most pressing problems associated with climate change. On a three-screen display opposite, these solutions are presented for the benefit of onlookers in a dashboard format (figures 3.8 and 3.9). On the central screen, territories and cities including Dubai, San Francisco, and the Brazilian state of Rondônia are singled out as "regions of change," accompanied by climate, demographic, and environmental data—such as the endangered species count for the region and the amount of CO_2 emitted per capita. To the left, recent satellite imagery paints the wall in familiar patches of green, blue, and beige. To the right, the same territory appears in dramatically altered form, projecting

FIGURE 3.8

Tega Brain, Julian Oliver, and Bengt Sjölén, *Asunder*, 2019. Courtesy of the artists.

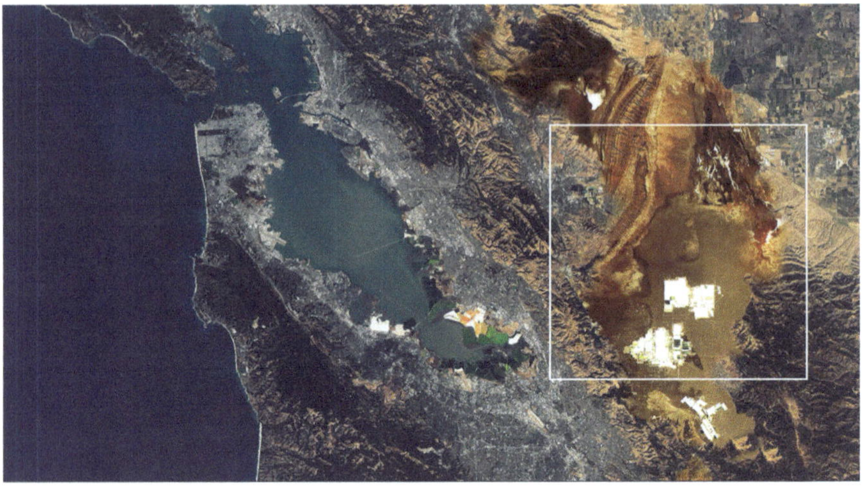

viewers into an imagined future in which planetary-scale geoengineering proposals have been realized. The computer's suggestions range from straightening entire coastlines and damming and redirecting rivers to uprooting urban centers with populations running into the millions.

The majority of the proposals are wildly unrealistic, even by the standards of some geoengineering schemes, but the fact that this is an artwork, created by Tega Brain, Julian Oliver, and Bengt Sjölen, titled *Asunder* (2019), means that the proposals were never intended to operate as concrete proposals to policymakers. Working with satellite imagery sourced from Landsat, the purpose-built software uses a generative adversarial network (GAN) to cycle through a series of future scenarios wholly determined by tech-centered solutions to climate change, generating "landscape edits" that embody the same aesthetic as the original source images. Despite the frequently absurd nature of the suggestions thrown up by

FIGURE 3.9

Tega Brain, Julian Oliver, and Bengt Sjölén, *Asunder*, 2019. Courtesy of the artists.

Asunder, it nevertheless represents the logical extension of many existing research projects that seek to bring together the fields of environmental management, artificial intelligence, and climate modeling. For example, researchers at Harvard and the University of Maryland have sketched out the conceptual design of a "wildness creator" powered by machine learning software that would "take on its own sentience and create a wild ecological space beyond any human control, and even human conception."[75] At the less hypothetical end of the spectrum, Anne Thessen provides an excellent survey of some of the ways AI has been used in ecology and the earth sciences, from modeling biodiversity and species distributions to predicting the ability of plant communities to adapt to new climates—a list that has grown significantly since the article was published.[76] In relation to geoengineering in particular, Microsoft and Nvidia have developed an AI tool designed to improve CCS, various simulations have been run in order to model the likely effects of solar radiation management, and work undertaken by Ben Kravitz to compare the predictions of various geoengineering climate models bears a striking resemblance to *Asunder*'s modus operandi.[77]

In much of this last body of AI research and development, metaphors of a machinic earth abound. Publications on geoengineering variously figure techniques such as SRM as "tools" to "fix" a broken climate system, adjust a "global thermostat," or "turn down the sun" as if it were attached to a giant dimmer switch—an addition that could be incorporated into one of the Odums' circuit diagrams without much difficultly.[78] These metaphors, while seemingly trivial, once again suggest the presence of a macroscopic view at work formulating an understanding of the planet as an engineerable entity. In an essay that links the legacy of systems ecology to the more recent datafication of the environment, one of the artists responsible for *Asunder*, Tega Brain, has argued that "even with the growing capacity of contemporary computing, it is dangerous to simply assume that more data equals more reality."[79] Vast territories of experience are irreducible to digits, and those ecological dynamics that can be quantified require acts of translation that inevitably embed certain worldviews and exclude others.[80] In sharp contrast to the use of satellite imagery in the work of Peter Fend, Brain explains in an interview that part of the provocation of the piece is to challenge the presumed objectivity of images produced in this way.[81] Satellite imagery is often framed as data visualization, rather than photography, the evidentiary status of the latter having been repeatedly challenged even before the era of digital photo editing and deep fakes. Fantasies of the impartiality of AI have also been thoroughly dismantled in recent years, with embedded bias revealed in applications ranging from airport security to advertising. The artists responsible for *Asunder* add environmental management software to the list of applications in which claims for the political neutrality of autonomous decision-making processes require closer scrutiny.[82] To take a phrase from the promotional video for the artwork, "*Asunder* positions both landscape and ecosystem as computational surface," and in so doing demonstrates the distortions necessary to "format" the world accordingly. As the media philosopher Jussi Parikka writes, formatting can work the other way around too, not only requiring particular media formats to capture portions of the world but also mobilizing images "as instruments of formatting environments, territories, landscapes, subjectivities, and social relations."[83]

The way in which ecologies are parsed as datafied artifacts plays a significant role in this formatting activity, a process that goes far beyond the terrain of machine learning. In an essay Brain wrote preceding the realization of the *Asunder* project, she links the recent use of AI in environmental management to the birth of scientific forestry and the cybernetic ecology of the Odums. This systems-based view posits plant and animal communities as knowable, bounded units comprised of discrete components "operating in chains of cause and effect," which, in turn, "invokes possibilities of manipulation and control."[84] This, I would argue, is also an assumption that lies at the very core of geoengineering as a practice. More than just an abstract logic, the work of forming ecosystems into discrete units—what Brockington and others call the "iconification of nature"—is inherently visual, bound up in a macroscopic view that sees wholes as collections of components and linear causal links rather than shifting webs of interdependency.[85] This is not just a question of resolution—of producing increasingly granular models to simulate the effects of geoengineering proposals—but rather of changing some of the underlying architecture upon which such models rely. In the case of atmospheric simulations, for example, this would mean shifting from a model based on carving the world up into multiple, discrete grids that are then stitched together, to working with models that are continuous across different and interconnected regions and weather systems, or, at the very least, models that facilitate greater potential for contiguity.

The assumption of linear causality also facilitates the smooth passage from explaining an ecosystem to using the resulting abstraction to predict and control its future development. Like all machine learning systems that rely on training datasets, if taken at face value, *Asunder* embodies the idea that with enough data, knowledge of environmental futures can be extrapolated from the facts of the past. As Brain makes clear, "the limitation of machine learning is that it can't imagine a world outside of its data set"—the task of overcoming this limitation and tuning models to the wider world is generally left to humans.[86] This is somewhat complicated by newer-generation AI and its capacity to "hallucinate" in ways that frequently defy expectation, although the latter represents more of a glitch than a paradigm shift toward a more abductive logic.[87] Such limitations are nevertheless laid bare in the spatial layout of the three screens. As the eye travels from left to right, it also passes through different temporal frames, from real-world satellite imagery drawn from the recent past to a simulation of possible futures. The farcical nature of some of the end results not only casts doubt on the ability of AI to grapple with problems of such magnitude without human oversight; it also casts a shadow back into the past, and onto the way satellite imagery is freighted with baggage from the very outset.[88]

The various AI-generated proposals occupying the middle screen of the installation have the simultaneous effect of removing the viewer from the decision-making activities of an autonomous system and resituating them in the position of an imaginary all-seeing climate engineer whose gaze surveys both the machine and the landscape upon which it looks. The dashboard aesthetic is key here, inspired by discussions of the "smart city dashboards" that promote data visualizations of a city's performance in real time. In both cases, delegated decision-making is reinforced by the power of oversight—in the literal sense of a macroscopic view "over" territory formatted as computational surface.

Asunder acts as both a prototype of technologies that are being discussed in earnest elsewhere and an immanent critique of the very same technologies made in a way that is typically excluded from the prototyping process. It is a work of sophisticated computer engineering that unpicks the idea that planetary systems can be engineered. Giving a name to this mode of practice, Julian Oliver builds on a long history of tactical media art by stressing the importance of "techno-political literacy" in his "Critical Engineering Manifesto," written collaboratively with Gordan Savičić and Danja Vasiliev.[89] In this pithy text, the familiar idea of engineering as the one-way imposition of control on a static object is unpicked, in such a way that "each work of engineering engineers its user, proportional to that user's dependency upon it." When the idea of a "user" is widened beyond the human, the stakes of revealing this dependency through acts of critical engineering grow even greater.[90] Instead of linear chains of cause and effect, an ecological understanding of the world posits moments of mutual transformation as commonplace, whereby collateral effects sometimes trigger "catastrophic" transitions from one stable state to another, and the spaces between seemingly discrete entities are where much of the action takes place.[91]

In a parallel discussion to that of critical engineering, Brain has spoken about the idea of "eccentric engineering," exploiting the etymology of "eccentric"—from the Greek *ekkentros* meaning "out of center"—to foreground the role of more-than-humans in engineering practices. Citing the example of a leaky water pipe, whose engineering success is unambiguously flawed from a human perspective, Brain argues that when viewed from the vantage point of eccentric engineering, the pipe could have demonstrable success in helping plants to thrive around the site of its leak.[92] Here, the word "engineering" is an awkward fit to describe what is essentially an unintended consequence, and I would suggest it only really applies when the leak triggers reflections or revisions on the intended purpose of the engineered object. One way of doing this, Brain argues, is to focus less on infrastructures hidden behind, below, or embedded within other structures, and more on "inter-structures" that lie in between, mediating relationships in often surprising ways, and integrate this into engineering practice. While works of critical engineering have been effective in revealing the evolving technological infrastructures that shape contemporary life under surveillance capitalism, for Brain it can also have a disempowering effect, reinforcing the unassailable status of big tech, rather than revealing its malleability or proposing alternatives.[93] By contrast, eccentric engineering uses critique as the warp around which to "weave in some kind of generative glimpse of a way to respond."[94] In doing so, it pushes at the threshold of what can be considered engineering, structuring relationships while also allowing room for unintended effects.

When looked at from this "eccentric" perspective, the power of geoengineering is to wrestle control back to a presumed center. Not only do many forms of geoengineering presume effective centralized political planning, but at a more fundamental level they rely upon clear chains of cause and effect in which collateral damage and unintended side effects are overcome by the triumph of humanity's technological ingenuity.[95] Whether by embracing the concept of eccentric engineering or through alternative conceptual frameworks, art has a role to play in emphasizing that environmental futures are always more-than-human affairs, in which other species have the capacity to participate as active agents in the shaping of our mutually imbricated worlds.[96]

ENGINEERING FOR A CHANGING LANDSCAPE

Each in different ways, Brain and Oliver argue that engineering can be more than a discipline of explanation, prediction, and control. In another corner of the world, two designers with a grounding in landscape architecture and planning have demonstrated that it can also be a practice of negotiation, in which engineering solutions actively accommodate uncertainty rather than seek to impose a rigid plan on the world.

Titled *SOAK*, the project is a direct response to severe flooding in the Indian state of Maharashtra in 2005, during which storms caused 944 millimeters of rainfall on the densely packed city of Mumbai in a single day. Millions of buildings were inundated with water, transport networks ground to a halt, and 1,094 people lost their lives—a figure that would no doubt be higher if the health impacts of contaminated drinking water were included in the count. As Neil Smith famously wrote in the wake of Hurricane Katrina, "There's no such thing as a natural disaster."[97] Rather, he explains, there are only natural events; it is the way in which events are prepared for and responded to that transforms them into disasters. When devastation is unleashed on human populations, it invariably falls upon steep social gradients that leave some communities disenfranchised and others relatively unscathed. Smith's quote, now so frequently used as to have become something of a catchphrase, acquires another inflection in the context of climate change, which has contributed to the increased frequency and severity of fires, floods, and droughts, calling into question the basis upon which these events can be considered natural in the first place.

While the precipitation levels of July 26, 2005, were the highest to hit Mumbai since records began, they were by no means anomalous. Extreme rainfall events have become increasingly common on the Indian subcontinent in recent years, with many scientists linking the shift to climate change.[98] Numerous factors made this flood worse than many previous ones, with blame apportioned to the city's unplanned expansion, poorly maintained drainage system, and construction on low-lying land reclaimed from the sea. For the urban planners and administrators tasked with safeguarding the city against future flooding, the solution was relatively straightforward: bigger and better drains, and the clearance of debris from the existing *nullahs*—the Hindustani name for wide canal-like watercourses designed to channel stormwater. The fact that, fifteen years later, floods in Mumbai are still a persistent problem is testament to the flaws in this strategy as much as its poor execution.[99]

For Anuradha Mathur and Dilip da Cunha, the answer is a great deal more complex. Instead of waging a war against water, molding it to the will of the city through concrete channels, they ask what it would mean to restore fluidity to a terrain that is inherently resistant to neat boundaries between land and water. Mumbai is a city that has attempted to defeat water on two fronts: on the one hand, against the monsoon rains, falling like sheets from the sky in sudden deluges, and on the other, against the sea, which laps at the shores of an estuarine landscape and enters the fissures of underground aquifers that are known to provide brackish water miles away from the coastline.[100] The flash point of the 2005 flood was a 18-kilometer stretch of water known as the Mithi River, which was overwhelmed when the downpour coincided with a high tide that pushed seawater up through the Mahim creek that carries the Mithi to the Arabian Sea. For Mathur and da Cunha, it is crucial to recognize this double aspect of the Mithi: It is not just a river that flows in one direction toward the sea; "it is Mumbai's estuary, a third coast," which thus "carries the possibility of a different reading of Mumbai, its past and its future."[101]

Rivers are typically represented on maps as a blue line, which splits in two when the river exceeds a certain width. It is a fixed entity, separated from dry land by banks that supposedly remain in the same position over time. While this may hold true in the short term for some rivers, Mathur and da Cunha argue that for the Mithi, it is a dangerous conceit. Elsewhere, in an expansive solo-authored work of historical geography, da Cunha has gone further still, suggesting that rivers are principally human artifacts and the line endowed with the power to separate land and water is a colonial invention that took centuries to institutionalize.[102] In this solo-authored work and in a joint exhibition at Mumbai's National Gallery of Modern Art in 2009, later catalogued in an accompanying book, the designers painstakingly unpick the history of how water has been confined and calibrated through cartographic fictions. The Mithi is a seasonal river, meaning it swells at some times of the year (figure 3.10). Although it is fed by the discharge of two reservoirs to the north, during monsoon season, Mathur and de Cunha argue, its "source is literally everywhere," and again, "the very notion of a line is called into question."[103] The same could be said for the border between land and sea, which shifts according to a cosmic timetable of tides and currents. Focusing in on a series of surveys produced by colonial-era mapmakers such as Captain Thomas Dickenson of the Bombay Corps of Engineers, Mathur and da Cunha show how these invariably ignored such tidal and seasonal fluctuations in the interest of producing a neat divide between land and sea. These were maps of a "fair-weather landscape," which with few exceptions excluded boundary-blurring phenomena such as swamps, mudflats, and mangrove forests—phenomena that are now the focus of increased global attention as coastal areas and littoral zones feel the effects of climate change.

Although Mathur and da Cunha do not themselves frame it in these terms, the cartographic fictions of Mumbai produced during the colonial era are an exemplary embodiment of the macroscopic view. While they were sometimes produced with striking levels of detail—a survey conducted between 1862 and 1875 claimed to show "every fixed object of the size of a foot square"—they nevertheless depicted a static landscape with

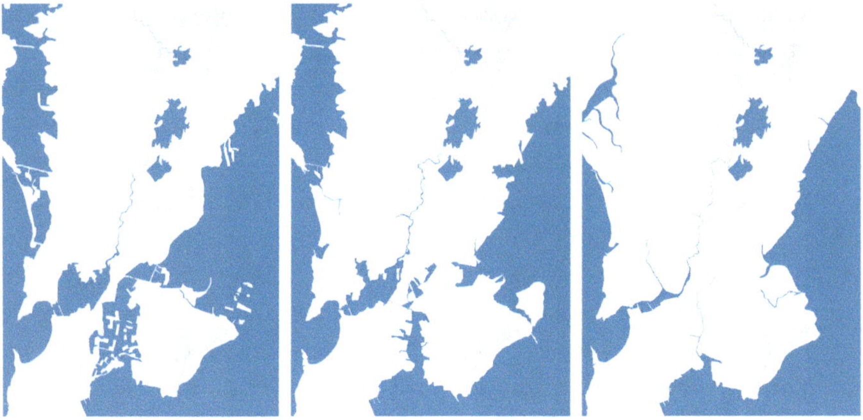

no temporal dimension.[104] The operative word here is "fixed," insofar as maps of this kind are constitutively incapable of capturing a landscape characterized by flux. But the word "object" is important too, speaking to the ways in which relations fall through the cracks of these representational schemas in their focus on discrete entities. In an era of sophisticated remote sensing and computational tidal modeling, one might expect such approaches to have been relegated to history, but as Mathur and da Cunha show, "their conceived "truth" has become the taken for granted ground of Mumbai—the basis of describing place and designing its future."[105]

More than just a passive representational grid laid upon the world, the macroscopic view is an active force within the landscape, endowed with the ability to bend it to its image. This power does not simply evaporate when such representational or discursive conventions are overturned. As climate change forces many cities to reexamine their cultural

FIGURE 3.10

Anuradha Mathur and Dilip da Cunha, *Making the Mithi*, 2009, appearing in *SOAK: Mumbai in an Estuary* (Rupa, 2009). Courtesy of the artists.

relationships to the borders between land and water, changing attitudes nevertheless still come up against the hard materiality of entrenched structures and processes. As bearers of sunken capital costs or substrates for systems that depend on their continued existence, the material objects that constitute the fabric of modern cities are often locked into paths that persist for many generations to come. This history of urban development is replete with examples of lock-in, but so too is the world of visual culture, which has established deep-rooted conventions that continue to shape environmental futures.

SOAK is not a project about geoengineering, nor indeed a project explicitly about climate change. In fact, some criticism of the work has complained that it doesn't devote enough attention to the inevitable destruction that sea level rise will bring to Mumbai's already fraught relationship to its waterways and intertidal zones, and that many of the existing plans developed by urban authorities—including those around stormwater management, green infrastructure, and "sponge cities"—are ignored.[106] Mathur and da Cunha's design proposals are large-scale, consisting of twelve suggested interventions that range from the construction of "nullah crossings" to the building of rippled monsoon surfaces and creek forts fringed by mangrove parks and serviced by sewage processing barges. However, these proposals do not come close to operating at the same scale as SRM or larger plans for CDR, nor to meeting the Royal Society's definition of geoengineering as "the deliberate and large-scale intervention in the Earth's climactic system with the aim of reducing global warming." Nevertheless, *SOAK* is a project that points to another mode of viewing the landscape, shifting away from the macroscopic view and toward something more dynamic, situated, and experiential, which could also harbor another mode of relationship to futures.

Accompanying Mathur and da Cunha's critiques of the colonial cartographic enterprise is a series of attempts to cultivate a different visual relationship to Mumbai's aqueous terrain. Strip collages, photo walks, and layered drawings populate the exhibition and accompanying book. Running through these diverse materials is a series of section views, which serve as a kind of visual motif for the project as a whole. Take figure 3.11, for example, which overlays two photos of very different meeting points between land and sea over an old map of the city: one, the concrete

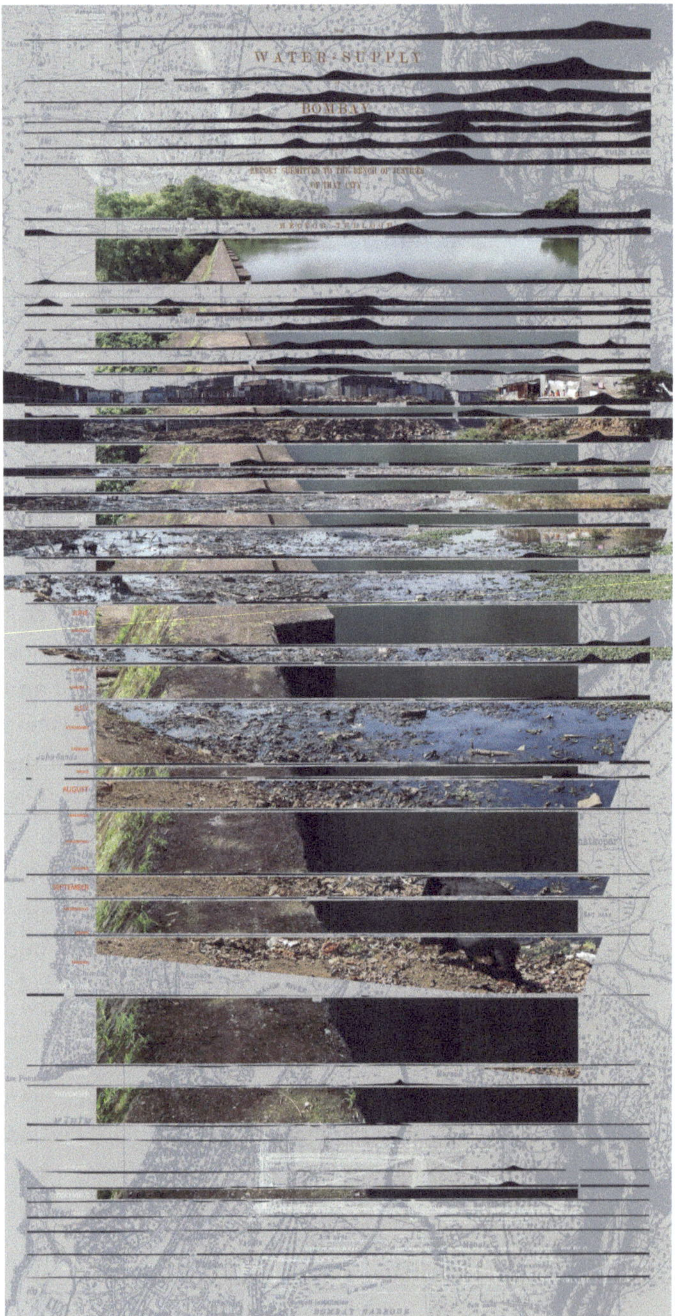
WATER SUPPLY
BOMBAY
REPORT SUBMITTED TO THE BENCH OF JUSTICES
OF THAT CITY

embodiment of a hard border dictated by the plan view; the other, an intertidal zone of flux and negotiation. A sequence of silhouetted section views runs from top to bottom, showing the elevation of the landscape in relation, rather than opposition, to the sea. It would be easy to read this composite image as a clash of two opposing views—hard versus soft, plan versus section—but in fact, its success is to facilitate reorientation from one to the other. By cutting the images together, the overriding aesthetic is one of gradients, where the plan view is swamped by the information it excludes.

Mathur and da Cunha's preference for sections, and especially those drawn in sequence can be attributed to their ability to register gradients and shift attention away from hard boundaries, but it also indicates an appreciation of the rhythms of the landscape. These are the rhythms of the sea as it swells and retreats, saturating the land rather than sitting passively on its surface.[107] Sectional views also enable recognition of the seasonal rhythms of the monsoon, whose waters can be carefully steered along fields and terraces rather than contained in two-dimensional blue lines. In other words, sectional views introduce a temporal dimension to the spatiality of traditional cartographic practices, and when laid in sequence, they transport their audience back to a world "rich in distinguishable moments rather than measurable distances"—a world once appreciated by seafarers, who, for Mathur and da Cunha, rarely take a position above the geographical spaces they traverse.[108]

What if the visual culture of geoengineering were to undergo a similar shift away from the macroscopic view? Instead of seeing the planet as a giant bio-geophysical machine whose workings are amenable to technological intervention, perhaps an alternative view would encourage acts of negotiation that accommodate change. If the speculative infrastructure

FIGURE 3.11

Anuradha Mathur and Dilip da Cunha, *Saturation*, 2009, appearing in *SOAK: Mumbai in an Estuary* (Rupa, 2009), 78. Courtesy of the artists.

of geoengineering is underpinned by "master plans," with roots in cybernetic ecologies of prediction and control, then perhaps *SOAK* points the way to a different mode of speculation in which preparedness is about having the agility to accommodate events that exceed the expectations of probabilistic futures-thinking. This doesn't necessarily entail adding to the innumerable "sacrifice zones," whereby environmental degradation is concentrated in some places as a means to protect others, but extrapolating outwards from these very territories, so that the sacrifice zone becomes the point of departure rather than an afterthought in engineering environmental futures.[109]

FAIR-WEATHER PREDICTIONS

Working in very different contexts and mediums, the artists and designers discussed in this chapter allow us to return to the opening questions with new insights: What does it mean to engineer climate futures into existence, how might engineering be separated from practices and systems that have had unintended effects on the climate, and what knowledge infrastructures and visual cultures allow planetary systems to be conceived as engineerable objects in the first place?

Mathur and da Cunha's proposition that modern attempts to control the flow of water in Mumbai are the product of cartographic fictions is, in many ways, a good metaphor for the tensions inherent in geoengineering, understood as a diverse collection of scientific, military, and policy proposals. In both instances, a process of abstraction is at play that bends the world to its image. In the case of geoengineering, the "fair weather landscape" that ignores monsoons and intertidal zones in the interest of producing stable representations is a landscape that covers the entire globe. Conceiving of the climate crisis as an engineering problem first of all requires a kind of clean-up operation, whereby the messy particulars of the planet are simplified in order to produce something with *a priori* engineerability. The particular is subsumed by the general, sometimes stretching its parameters, but rarely challenging its principles at a fundamental level.[110] The confidence of the geoengineer as an ideal figure is one result of this process, a subject position derivative of the climate model, albeit without the necessary tuning of such models to reality seen at full resolution.

While this chapter has adopted a critical perspective on geoengineering, it is worth underlining that many of the abstractions upon which it relies are nevertheless indispensable to modern science, and without them, climate knowledge would be robbed of one of its central means of apprehending the planet. Even relatively localized phenomenon such as algae blooms in the North Sea or the increased frequency of extreme weather events in Mumbai would be impossible to understand without modeling the larger systems of which these phenomena are a part. And while ever more datafication of the world does not automatically equate to more understanding, the macroscopic view does enable partial knowledge upon which governments and individuals alike have come to rely. For Tsing, global models are also resources insofar as they can facilitate collaboration, with generalizations smoothing the path toward negotiation by "making its necessity disappear into a seamless globe."[111] Problems arise when partial knowledge is treated as if it were complete knowledge, and the seamless globe of the model becomes a substitute for the messy particulars of the planet. In stark contrast to Moiseyev's idea of steering planetary environmental governance through "negative guidance," many proponents of geoengineering translate partial knowledge into prescriptive solutions. This acts to obfuscate the onto-epistemological gap between the planet and its representation, underpinned by an extraordinary faith that present-day models are sophisticated enough not to be overrun by real-world events. In reality, many of these models are unable to accurately simulate the atmospheric effects of natural phenomena such as volcanic eruptions, which have been observed by humans long before the effects of greenhouse gasses.[112]

Where novelty and the unexpected *are* accommodated by geoengineering proposals, it is often accompanied by an implicit assumption that history is only ever made by humanity, and a small section of humanity at that (the infamous epigraph of the *Whole Earth Catalogue*, mentioned in the previous chapter, "We are as gods, and we might as well get good at it," acting as an emblematic expression of this anthropocentrism).[113] At the core of proposals to engineer the climate is the question of intent. Geoengineering did not begin "soon after our species started using fire for cooking," as Lovelock once claimed, for the simple reason that having an impact on one's environment is not the same as attempting to engineer it—here, intent makes the difference.[114] These attempts are not accidental or unconnected, as the cursory summary of geoengineering in this chapter has attempted to show. They are underpinned by a growing knowledge infrastructure composed of funding streams, patent applications, scientific expertise, and emerging forms of environmental governance that narrow the analytical parameters of "wicked" problems to those with technical solutions. With the question of intent, this infrastructure harbors within it a specific mode of relation to the futures.

And yet, perhaps the question of intent can be dislodged ever so slightly from its position at the heart of geoengineering. As a form of futuring, or speculative infrastructure, perhaps engineering has the capacity to be less about imposing master plans and instead come to expand the range of tolerances and flexibilities already built into some of its practices.[115] The examples in the second half of this chapter have been selected not because they propose compelling solutions to climate change or biodiversity loss but because they reveal ways in which engineering can further loosen its addiction to control, coming to accommodate unintended side effects or moments of mutual transformation at the meeting points of technology, visual culture, and ecology. This is the threshold at which engineering futures starts to morph into other speculative practices, ones that recognize the future-making power of more-than-human agents and the vast swaths of reality that defy systematization. In their capacity to interrogate systems and histories of representing the environment, art and creative practice can suggest ways in which predictive reason might be subtracted from engineering practices, ultimately undermining the confidence necessary for geoengineering proposals to be translated into prescriptive solutions to climate change in the first place.

4

ENACTING

Alternative economies are not then just microcosms or prefigurations of a degrowth world. They are incubators, where people perform every day the alternative world they would like to construct, its logic rendered common sense.
—Giorgos Kallis, *Degrowth*[1]

Degrowth's call is not for a return to a past that never existed, but for a simultaneous production of the present by the past and the future.
—Giorgos Kallis and Hug March, "Imaginaries of Hope: The Utopianism of Degrowth"[2]

INTRODUCTION

This chapter begins with a piece of political theatre that took place in France in the early 2000s. In contrast to some of the projects discussed in the previous chapter that seek to intervene on a planetary scale, this action privileged intimacy, direct action, and influence outside of the major knowledge infrastructures of climate science and policy. And yet, despite this limited sphere of influence, the action speaks to a much wider context, revealing the contours of a debate around environmental futurity and economic growth that has since found its way into parliamentary politics and mainstream discussions of climate adaptation and mitigation.[3]

The action began in a small village in the south of France in 2006, where the political campaigner François Schneider embarked on a meandering journey from village to village with his donkey, Jujube. In total they walked around 1,500 kilometers before concluding their trek at the town of Magny-Cours in the Nièvre region of Burgundy. Over the course of twelve months, Schneider and Jubube picked up around five hundred walking companions, all rallying around the campaign's central message of *décroissance* or "degrowth"—a term that is more than a slogan for Schneider, and represents a "concrete, rational process aimed at reducing the consumption of natural resources at a global level."[4] The national press covered much of Schneider's journey with Jubube, which took in numerous symbols of rural decline: former vineyards with rotting vines, scrapheaps filled with piled-up cars, and a traditional bread oven that had lain unused for fifty years in the town of Bessenay. "That night," Schneider narrated in his account of the journey, "the bread oven is restarted to feed the marchers." A baker associated with the local group the Hive of Ecology made what Schneider describes as "the bread of degrowth" using renewable energy and organic wheat, and in doing so performed a vision of an environmental future characterized by bioregionalism and conviviality.[5] The march also took in the Formula 1 Grand Prix track—Magny-Cours' largest tourist attraction and, for Schneider, "a symbol of our deadly growth society"—providing a contrast to the plodding of a pack animal as it inched its way across the country. Rich in symbolism, the choice to travel via donkey might suggest that ideas around degrowth hinge on a fundamental privileging of slowness over speed, or tradition over modernity, and yet, as this chapter aims to show, the temporal politics around which the movement turns cannot be diagrammed in such a linear fashion. While there's no denying that the political action exuded nostalgia for a kind of bucolic pastoralism, degrowth can nevertheless be viewed as an inherently speculative social movement, "a simultaneous production of the present by the past and the future," comprising practical explorations of what societies freed from the economic imperatives of GDP might look like in the present.[6]

Schneider's walking tour, while the first of its kind to promote degrowth in France, did not announce the arrival of an entirely new concept on the French political landscape. It was built upon a history of critiques of economic growth stretching back at least to the early 1970s. On an international stage, the most famous of these is the Club of Rome's *Limits to Growth* report published in 1972. In France, a handful of publications by the likes of André Gorz, André Amar, and Nicholas Georgescu-Roegen cohered around the term *décroissance* in the same decade and helped to form the academic disciplines of ecological economics and political ecology.[7] By the time of Schneider's walking tour, the foundations of a wider European social movement had been laid, with protests advocating degrowth in Lyon (2001), Italy (2004), Catalonia and Spain (2006).[8] For its early adopters, degrowth served as a "missile word" that sought to challenge the assumed benevolence of economic growth and sustainable development, prompting discussion on how and where to impose limits on ever-expanding consumption. For its detractors, the culture that has arisen around degrowth is characterized by a "folk politics" that unhelpfully valorizes the local and small scale as a site of authenticity, and is ill-equipped to tackle a complex global economy in a meaningful way.[9] The very fact that such a debate is taking place is indicative of the ascendency of degrowth as an organizing pole for discussions of climate change mitigation and adaptation, as indicated by its inclusion in recent IPCC reports, where the term is evoked as a plausible basis for future scenarios.[10] Whether or not degrowth will continue to be accommodated within the scenarios, models, and other speculative infrastructure of mainstream climate science remains to be seen. Rather than focusing on the ways in which these accommodations have taken place thus far, this chapter instead attempts to unearth a mode of speculation within degrowth itself, as well as projects that could be considered adjacent to the movement. Where the preceding chapters explored some of the tools and techniques that seek to build bridges to futures that remain out of reach, this final chapter explores deliberate attempts to actualize—or "enact"—futures in the present, embodying an impulse within futures studies of the 1960s and 70s, "to create the future now . . . through action."[11]

Central here is an acknowledgement that degrowth is not only a "missile word" or body of heterodox economic thought; it is, as Schneider's walking tour suggests, part of a broader culture that harbors a specific set of speculative practices. Rather than producing representations or blueprints of utopian visions, many of the enactments that comprise this broader culture involve the practical construction of what Chris Carlsson calls "nowtopias": spaces where the social and technological foundations of post-capitalist forms of life are cultivated.[12] The aesthetic enters here not as a visual modality akin to the aerial or macroscopic views explored in previous chapters, but as a form of prototyping that challenges what many have come to expect from visions of environmental futures. Insofar as enactment involves bringing futures into the present, it also makes them more amenable to being "felt," and thus aesthetics also relates back to the realm of embodied sensations. The decision to open this chapter with an earthy story involving a donkey is, in this respect, intended to reorient the reader away from the techno-positivist sci-fi imaginaries that dominate representations of environmental futures toward narratives that weave different temporal and technological horizons. In the sections that follow, artworks and exhibitions come to embody such enactive practices in different ways, first of all from an exuberant artistic expression of the "growth paradigm" in an exhibition curated by Rem Koolhaas, and in subsequent sections in artworks by Adrian Balseca, Kathrin Böhm, and others that explore alternative economic infrastructures and ideologies, complicate linear timelines of past–present–future (particularly in regards to the spatial politics of the rural), and raise important questions of the scale and scalability of artworks that enact, rather than simply imagine or represent, environmental futures. To set the stage for these artistic explorations, the next section will first explore two facets of growth—one based upon linear temporal structures and accumulation and another more closely aligned with ecological thought and practice.

GROWTH SINGULAR PLURAL

What exactly is the target of the degrowth movement, and how might it be possible to distinguish between growth understood as a purely economic phenomenon, and forms of growth in an expanded ecological sense? At its most basic, growth is a quantitative or qualitative change that takes place over time, and one that relies upon a comparison between two or more moments in the life of a person, place, or thing. For this act of comparison to work, time has tended to be transformed from something multiple and diverse into a singular, stable backdrop against which growth can be measured.

Sociologists of time such as Barbara Adam have been persuasive in showing how this commonplace understanding of time as a stable and consistent backdrop to growth has a distinctly modern genealogy. In her influential book *Timescapes of Modernity*, Adam writes of the historical emergence of "industrial time," which she sees as being shaped by three major forces: "a) the invariable beat of the clock, b) the economic commodification of time and c) the scientific use of time as measure of abstract motion."[13] Tracing each of these influences to the birth of modernity, Adam shows how they are inextricably entangled, with clock time both facilitating the quantification of time as units of financial value and reinforcing the epistemological claims of Newtonian physics, which contains a range of temporal assumptions, including linear causality, reversibility, rationalization, and objectivity.[14] Adam's three-pronged analytic—which she abbreviates to "machine, economic and laboratory time"—leaves room for elaboration around the role of labor in the formation of industrial time, for example, in the corrosive effects that the creation of an urban proletariat had on agrarian communities' temporal relationships to the land in Western Europe, or the steamrolling of Indigenous temporal frameworks through colonial histories of enslavement and dislocation.[15] Despite these gaps, the account of industrial time nevertheless serves as a useful skeleton key to unlock some of the dynamics at play, as well as the way in which these dynamics become "habits of mind" that make it increasingly difficult to conceive of time in a way that isn't linear, abstract, and dependable.

Just as scholars have shown the hegemony of industrial time to be contingent on certain historical conditions, it is also possible to historically situate economic growth as a peculiarly modern preoccupation. Historian Gareth Dale speaks of the emergence of a "growth paradigm" in the seventeenth century, before which only scattered reflections existed on the idea of "the economy" as a discrete entity, let alone one that could be said to grow in size year upon year.[16] What Dale identifies as the growth paradigm does not begin with the quantification of annual economic growth as we know it today, but with a set of pre-conditions in which "the conceptualization of time as an abstract, infinite continuum" serves as an important pre-requisite.[17] Picking up momentum over the next three centuries, the growth paradigm finds expression in all areas of cultural and political activity, fueling colonial expansion and class stratification in the imperial core. It is only in the twentieth century, however, that it begins to be quantified at a macroeconomic scale, with the systematic accounting of GDP in the 1930s and the promotion of the measure internationally at the Bretton Woods conference a decade later.[18] What followed was the establishment of a global infrastructure for calculating market activity, with organizations such as The Organisation for Economic Co-operation and Development (OECD) and the World Bank setting growth targets and "harmonizing" economic policies across the world—a popular euphemism for the violent shock therapies that characterized the spread of neoliberalism in the postwar era.[19]

Despite the purity of its numerical expression in the form of GDP, it would be wrong to think of economic growth as an exclusively economic concern. As William Davies writes, it is merely a crude representation of a wider "sense of progress as a collective temporal experience," cultivated and reinforced by economic doctrines such as Keynesianism.[20] Nor would it be correct to assume that economic growth is a relatively constant feature of national economies that pre-exists the twentieth-century obsession with GDP—the numbers increasing year upon year without attracting attention as something worthy of measurement. As economists such as Angus Maddison and his team have shown, economic growth is a modern phenomenon not only in name but also when today's metrics are retrofitted onto the past. Prior to the modern period, growth in economic productivity rarely occurred without an increase in population.[21]

Degrowth, then, represents an attempt to step outside of the growth paradigm forged in the crucible of capitalism, colonialism, and the hegemony of industrial time. As many of its proponents are at pains to stress, it does not argue for an enforced contraction of the economy (the likes of which were all too keenly felt during the coronavirus pandemic), nor does it seek to impose economic restrictions on countries whose economies have been held back through asymmetrical power relations and the endurance of exploitative colonial economic regimes. While a decrease in the "throughput" of the economy (that is, a decrease in the energy, materials, and waste flows) is a long-term goal of the degrowth agenda, it is also important to stop thinking in terms of economic growth in the first place.[22] As Serge Latouche remarks, in this sense, the word itself is misleading: "We should be talking at the theoretical level of 'a-growth,' in the sense in which we speak of 'a-theism,' rather than 'de-growth.'"[23]

While the vast majority of national and supranational economic policies remain firmly attached to the ideology of economic growth, mainstream politicians with differing ideological allegiances have also entertained alternative metrics that quantify affects such as wellbeing and happiness in recent years.[24] Leaving aside the numerous and now well-established critiques of such alternative metrics, which are often accommodated with very little difficulty into existing frameworks of economic governance, it is worth contemplating what a form of growth might look like that is more closely aligned with ecological, rather than economic, imperatives. In *Timescapes of Modernity*, Adam contrasts the temporal assumptions embedded in economic growth with "the timescape of living processes of re/production and re/generation."[25] In contrast to narrowly conceived definitions of economic growth, ecological growth is both expansive and inescapable, "inextricably tied to the basic material conditions of earthly existence, to sun, air, water and soil."[26] In this expanded sense, Adam contends, growth need not be wedded to the modern phenomenon of industrial time. Insofar as it operates irrespective of context, she argues, the abstraction of industrial time stands in a "problematic relationship to the contextual, irreversible temporalities of life and the multiple rhythmicities of nature."[27] Here, the seemingly monolithic abstraction of growth splinters into diverse processes underpinned by a heterogeneous set of temporal structures, a point reinforced by more recent scholarship that reclaims growth from its dominant economic definition.[28] No longer simply an aggregate measure of ever-expanding economic activity, in these accounts, growth parts ways with familiar notions of progress and the collective temporal experience it implies.

Indeed, broader considerations of ecological growth require making the temporal experience of life beyond the human an active matter of concern, following the example set by scholars of multispecies communities in recent decades, such as Michelle Bastian and Deborah Bird Rose.[29] In a compelling example of the kinds of embodied temporalities that can be found in ecosystems, Rose considers the co-evolutionary coupling of *Pteropus* (commonly known as flying foxes) and myrtaceous trees in Northern Australia. In doing so, she shows not only how this relationship is mutually beneficial to both species but also how the exchange hinges on crucial sequences and synchronies in how the species interact with one another. The trees that provide nourishment for the fox produce flowers in a sequence that lasts much of the year, in a movement from higher and drier parts of the landscape to the lower banks of the Victoria River. The foxes arrive at just the right time to profit from the beginning of this display, a synchrony that enables them to follow the movement of the blossom as it spreads through the territory, pollinating plants and spreading seeds along the way. The combination of these encounters for Rose is a "multispecies knot of ethical time," where sequences and synchronies intersect to form "a temporal site of embodied interface."[30] And yet, in a changing climate, these knots are starting to come untied. With the diversity and distribution of forests in decline, flying fox numbers have also plummeted, beginning a spiral of "functional extinction" characterized by a "loss of connectivity and mutuality."[31] Thinking back to the systems ecology discussed in the preceding chapter, the issue is not just the loss of one node in a fixed network but also species losing their footing in each other's temporal worlds, which further reduces the ability of the community as a whole to flourish in the face of change. The growth and decline of these species are almost impossible to comprehend through the lens of disembodied industrial time, which is geared toward the measurement of discrete units rather than complex webs of interdependency. Taken at a low temporal resolution, and paying attention to individual species of myrtles, the tree numbers might indicate that they are far from becoming extinct. And yet, when the sequence and synchronizations of their relations with other species are considered, a very different story unfolds.

This is a story that extends beyond extinction narratives—functional or otherwise—in the context of climate change. In the past few decades, a quiet revolution has taken place in the environmental sciences, as nonlinear accounts of ecosystemic change have come to challenge once dominant ideas founded on linear time such "vegetative succession."[32] Whereas it was once thought that ecological communities such as forests changed much like an individual organism, passing through a series of more or less predictable life stages before eventually reaching a stable "climax community," today there is greater awareness of so-called "catastrophic transitions" between two or more stable states that may result from a change in environmental conditions.[33] In many cases it is not possible to simply rewind the clock and restore the conditions that gave rise to the first ecological community—for example, restoring soil nutrient levels as a means to reverse desertification—for there is no guarantee that any given ecological community will arise under a particular set of environmental conditions. The possible forms that an ecological community may take under a given set of conditions are multiple, and so too are the changes between them. Here it might be tempting to reach back to chapter 1 and try to understand the multiple trajectories of ecological growth through a scenario framework, with each possible future of a given ecosystem fanning out from the present. However, scenario thinking still presupposes a certain degree of linear development, as set out earlier, and therefore has limited analytic potential in this case. Environmental scientists have instead appealed to the theory of dynamical systems, according to which an ecological community can be likened to a ball on an uneven surface, being pulled toward various possibilities that serve as "attractors." Here, there is no forward and no backward on a path toward an ecosystem's climax community. Futures of ecological growth instead represent the complex push and pull of different conditions, some of which are external (as is the case of interannual variability of weather systems) and some of which are created by the ecological community itself.[34]

This handful of examples paints a picture of ecological growth as a phenomenon in which nonlinear temporalities are widespread, perhaps even ubiquitous.[35] This is not to say that more linear forms of growth are entirely absent in more-than-human ecologies, or that it cannot sometimes be measured against the disembodied standard of industrial time, as demonstrated by the metric of "growing degree days" used in commercial agriculture.[36] What is worth underlining, however, is that whilst industrial time is singular and universal, the temporalities of more-than-human ecologies are multiple and diverse. This leads to diverse forms of growth too, moments of flourishing indexed to a heterogeneous mix of rhythms, patterns, and synchronizations. Here, Adam's distinction between two different forms of growth could be expanded to encompass a greater range of possibilities, helping us to think at the level of "agrowth"

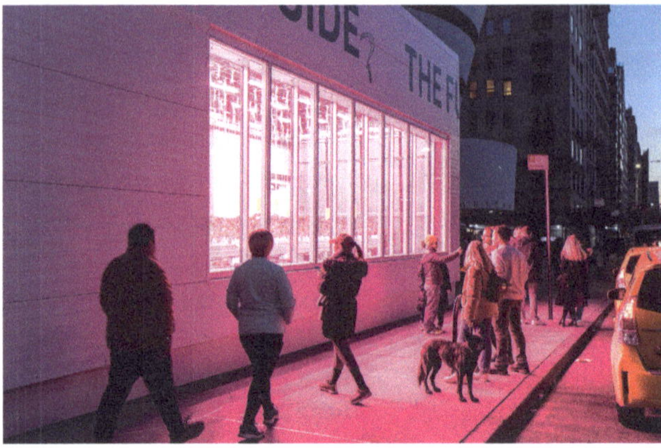

when it comes to the economy while also exploring more complex and relational forms of growth elsewhere.

PINKHOUSES AND PRECISION FARMING

While the preceding section described some of the temporal characteristics of economic and ecological growth, this section looks to a recent exhibition at the Guggenheim Museum in New York to think about how the two come into conflict in that vast territory of the planet known as "the countryside." In stark contrast to the convivial pastoralism of Schneider's walking tour, here efficiency and intensification are celebrated elements of an underlying ecomodernist sensibility. Curated by the architect Rem Koolhaas and his research group AMO, *Countryside, The Future* (2020–2021) advertised itself with pink LED bulbs spilling their light waves onto Fifth Avenue (figure 4.1). These comprised part of a window display that

FIGURE 4.1

Left: Digital photographs of OMA/AMO, *Countryside, The Future*, 2020. Photo by Laurian Ghinițoiu, courtesy OMA/AMO. Featuring a grow module by 80 Acres Farm.

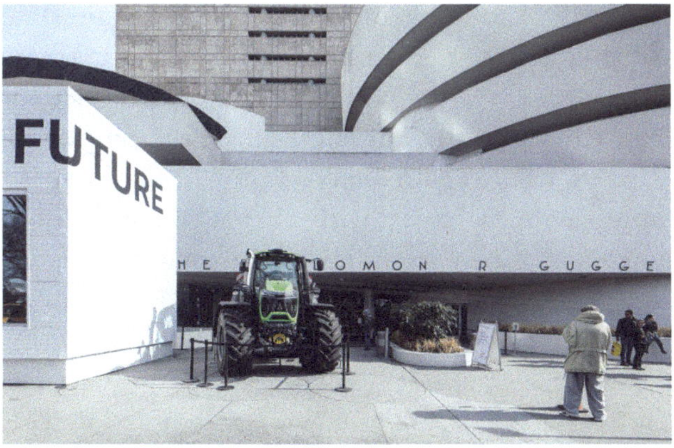

functioned as a tomato grow module produced by the organization 80 Acres Farms, whose indoor farms operate without sun, soil, or much of a human workforce. Alongside the grow module, a state-of-the-art Deutz-Fahr 9340 TTV Warrior tractor sits like a sentinel on the pavement outside the exhibition, its novelty, the visitor is later told, lying in its ability to be controlled remotely via an iPad (figure 4.2). Such tech-centered solutions are showcased throughout the exhibition and accompanying publication, with precision agriculture, rural e-commerce villages, and the posthuman architecture of rural logistics and fulfillment centers taking center stage. Speed, efficiency, growth, and expansion are everywhere on display and, in typical Koolhaasian fashion, presented with a breathless awe that risks brushing aside any disquiet or critique of the future worlds being showcased and brought closer to the present.

FIGURE 4.2

Right: Digital photographs of OMA/AMO, *Countryside, The Future*, 2020. Photo by Laurian Ghinițoiu, courtesy OMA/AMO. Featuring a Deutz-Fahr 9340 TTV Warrior tractor outside the Guggenheim Museum, New York.

There is little appetite for the kinds of direct, programmatic statements found in degrowth literature here either. Koolhaas and his team claim to have simply created a survey, or, as he puts it, a "pointillist portrait," of that vast expanse of the earth that falls outside the category of urban experience. Pursuing what the architect playfully calls "Martian strategy"—in short, approaching a subject with calculated naivety in order to expose elements that an expert might miss—the mood of the exhibition is for the most part interrogative rather than imperative.[37] And yet, as one critic pointed out, in terms of a viewing experience, the resulting exhibition consists of such an overabundance of text, images, and objects that it seems to actively discourage deeper modes of contemplation. This information dump, somewhat scaled back in the accompanying publication, instead rewards those "able to scan for connections and change channels, to switch from screen to screen with virtuosity," in a way that becomes indistinguishable from the mode of attention encouraged by online media.[38] Attempts to write about exhibitions of this sort often become somewhat impressionistic endeavors in themselves, seeking to capture the atmospherics of aggregated information, rather than gaining enough of a foothold to articulate a critical position. In an effort to avoid this mode of passive reconstruction, the intention here is to find moments in the exhibition that might function as parts that stand in for the whole, and better understand the modes of speculation the project puts to work in the process. The result is highly selective, but in an exhibition comprising several thousand images, the few that follow are nevertheless able to serve as metonyms that unlock some of the content of the project and its unspoken ideological commitments in a way that would not be possible by attempting to document it in full.

The first example is an experiment in "pixel farming" representative of the influences that have helped shape Koolhaas's understanding of the countryside as a geographical locus, concept, and field of activity. The second is a photomontage that featured in the exhibition and serves as a cover for the accompanying publication—already a metonym of sorts—that reveals a set of temporal assumptions underpinning the exhibition as a whole. Although global in scope, featuring case studies in Africa and Asia, the idea for the project was seeded in the Dutch countryside, where the increased visibility of the Netherlands' agro-scientific-industrial complex, driven by Wageningen University & Research (WUR), had given Koolhaas pause for reflection on his commute between Amsterdam and Rotterdam.[39] WUR has become a global center for innovation in food science, making the tiny country it calls home one of the biggest food exporters in the world.[40] This is also a country shaped by several hundred years of hydraulic engineering, where an intricate system of dikes and embankments mitigate the present and future risks of water encroachment. More recently, it has become a landscape of vast greenhouse complexes, some covering upwards of a hundred acres, within which climate is precision-engineered to maximize the yields of specific crops.[41] Diurnal rhythms cease to have much relevance here, and growth extends into the future in a predictable, linear fashion.

The exhibition features an experimental outgrowth of this industry in the form of a pilot project led by Lenora Ditzler on "pixel farming." By way of a definition of pixel farming, Ditzler compares it to other approaches in agriculture that encourage biodiversity such as strip cropping, but, "instead of one or two crops in rows, you saw clusters of different sizes, shapes, and colors daubed across the landscape."[42] Cultivating several different companion species in one "high resolution" plot, Ditzler and her team draw on non-intensive horticultural methods but attempt to expand them to an industrial scale. While stopping short of trying to replicate the kinds of nonlinear ecological dynamics discussed in the last section, here, growth is nevertheless a complex, multispecies project, which involves both sequential and synchronous time. In this way, pixel farming represents a conscious break from the orthodoxy of modern Dutch agriculture that relies on monoculture plantations and regular fertilizer inputs, and instead draws inspiration from agroecological approaches not dissimilar from the Indigenous farming techniques encountered in chapter 2.[43]

And yet, in its speculative dimensions, the project also echoes the techno-positivist sensibilities of intensive agriculture that has emerged from Holland's "Food Valley"—a name that makes a deliberate nod to Silicon Valley.[44] As Ditzler readily admits, for all its parallels to planting systems such as milpa, pixel farming is also heavily reliant on technology, using statistical methods and the expertise of plant modelers to simulate the expected growth of the plants in question in "virtual pixel plots" alongside actual planting trials. The aim is to produce a system that can scale, which presupposes the invention of machinery to sow, maintain, and harvest crops from large fields of this complexity. For this to happen, Ditzler envisages that "technological advances—and automation—will be necessary."[45] Part of the rationale for using the metaphor of the pixel to describe what are essentially 1.5 × 1.5–foot crop cells is to "make the idea of radical farming more palatable to digitally minded people," and more amenable to machine-orientated considerations of resolution and pattern. Pixel farming thus pushes at the limits of existing knowledge and technological capabilities, requiring significant levels of investment and innovation before its promise can be fulfilled. Its current relationship to futures is thus one structured around producing expectations, which, as numerous STS scholars have shown, are key to gathering the necessary alliances and resources to support the realization of the futures being promoted.[46] One such expectation is that in significantly increasing crop diversity and yields at scale, pixel farming would represent a vast improvement on much commercial agriculture as it is practiced in the West today.

A critical appraisal of pixel farming, however, might find it to be too reliant on the paradigm of agro-technological innovation that depends on continued economic growth and capital investment to move forward. For these reasons, the project could be aligned with an ecomodernist sensibility that places faith in technological innovation to solve environmental problems. Less a social movement than a loose set of ideological commitments to maintaining current living standards through increasingly intensified agriculture, nuclear power, and forms of geoengineering, ecomodernism represents a collection of proposals that are irreconcilable with a degrowth perspective and underpinned by a linear conception of technological progress.[47] Central among them is the idea that economic growth can be "decoupled" from environmental impacts, transforming itself into green growth that gradually becomes self-sustaining without the need for excessive resource extraction. Advocates of degrowth are rightly skeptical of this and call for a more radical shift in the metabolic basis of society. As Erik Gómez-Baggethun points out, decoupling, or "dematerialization" of GDP growth "has come true only in developed countries that have outsourced their industry to developing countries with cheaper labour force and softer environmental regulations."[48] When the true externalities of decoupling enter the picture, the ecomodernist faith in green growth comes to seem willfully naive—the product of an economic paradigm that is structurally blind to effects that don't fit on a balance sheet.

With its ecomodernist underpinnings and commitments to agroecology, pixel farming embodies a confluence of the two forms of growth set out in the previous section of this chapter. On the one hand, it embraces the nonlinear temporality of companion planting in all its sequential and synchronous complexity; on the other, it can only succeed against an implied backdrop of continued economic growth and innovation underpinned by expectations to solve environmental problems. It both represents an implicit critique of monocropping, intensive farming techniques that can have a disastrous effect on biodiversity and soil health, and reflects a deepening reliance on technological innovation that can only serve to increase the material throughput of the economy. Ultimately, in its privileging of new technological solutions without regard for the socioeconomic relations they produce, pixel farming resolves this tension by subsuming ongoing ecological growth within an assumed framework of economic growth and speculative capital investment.

An ecomodernist sensibility is further revealed in a second moment of the exhibition that serves as a useful point of orientation for understanding considerations of temporality. At the top of the Guggenheim's rotunda, a photomontage juxtaposes an image of Russian peasant farmers in 1909 with a photo of an indoor farm in the Netherlands run by the company Koppert Cress, once again featuring pink grow lights (figure 4.3). The images are overlaid with the words "then" and "now." In between them, a wall text speaks of the shift from the countryside as a "stable environment where everyone—man, woman, child—knows their place" to a landscape where life has become inorganic, with "no ground, no sun, no place, isolated figures, monochromatic, fully enclosed and proud of techniques and efficiency."[49]

The photomontage is unabashed in its generalizing tendencies, positioning the Russian countryside as an anchor point for class and gender identity that sits somewhere outside of historical time, reinforced by the use of the present continuous tense ("knows"). The narrative establishes a clear figure–ground relationship between the past and the present (and

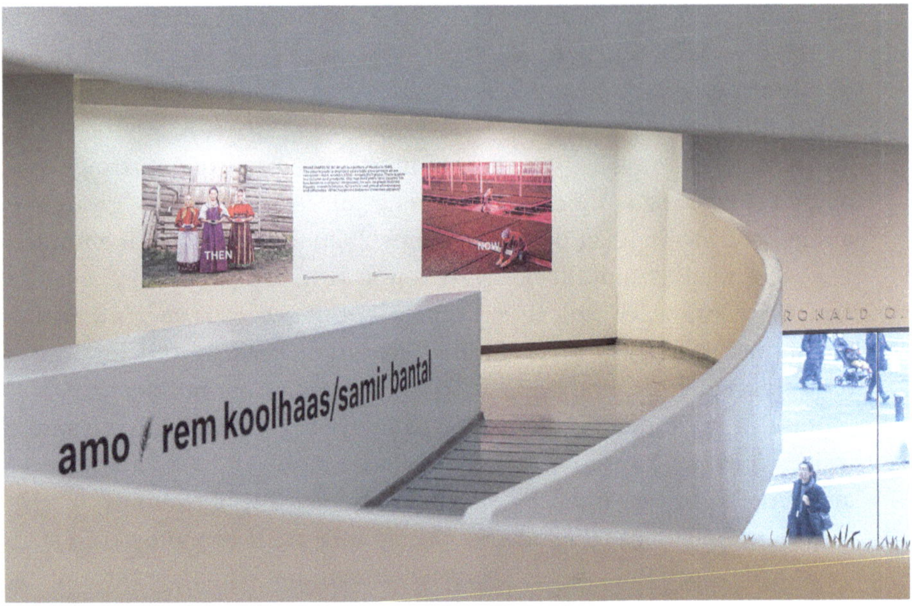

by extension, *the* future), with the former serving as a stable backdrop against which a modernist telos of progressive agricultural innovation reaches forward toward optimal efficiency. Within this temporal framework, a spatial politics also makes itself felt. The "now" of Koppert Cress is, once again, located in the Netherlands, whilst the rural past is always elsewhere—if not in Russia, then, by implication, in countries that have been slow to embrace the ag-tech and industrial farming techniques so prevalent in Koolhaas's home country. There are no attempts to provincialize this form of "development"—in the sense that it could be seen as just one, culturally specific form of agriculture—nor is there any acknowledgement that in some parts of the world, including in the Netherlands

FIGURE 4.3

Digital Photographs of OMA/AMO, *Countryside, The Future*, 2020. Photo by Laurian Ghinițoiu, courtesy of OMA/AMO.

itself, futures involve a rejection of the selective vision of rurality that the exhibition presents.[50] This is similarly evident in a conceptual outline for a collaboration between AMO and Volkswagen (VW) that seeks to develop an electric tractor and solar charging station designed for use in sub-Saharan Africa, also on display in the exhibition.[51] While there could be notable benefits from such technologies, there is no reflection on what sociopolitical relations would underpin its design process or delivery, other than an indication that the tractors would not be sold to individuals but rented out by VW, in effect creating yet another relationship of dependency on a Western company for the means to increase food production.[52]

As this last project demonstrates, and in regard to the exhibition's claims to have produced a "pointillist portrait" of the countryside, it becomes clear that this portrait has been made from a very particular angle. Its exuberant embrace of tools and technologies that rely on continued capitalist innovation serves as a prominent shop window to promote such projects in a major center of global power.[53] In the case of the collaboration between AMO and VW, "the future" of the countryside on show in the exhibition is one in which Koolhaas's business interests are actively tied up. To borrow the language of strategic foresight, this is not so much an exhibition that seeks to identify "weak signals" or subtle clues as to what futures might hold, but a collection of projects, partnerships, and marketing strategies that seek to enact a specific techno-capitalist imaginary of the countryside.[54] As Koolhaas himself puts it, architects are "not so much predicters of the future as enablers of the future."[55] (I would add, drawing inspiration from Tony Fry, that they can also be disenablers of futures, introducing objects and ideas into the world that negate possibilities).[56] The futures being enabled are aimed at dislodging the misguided assumption that the engines of progress and innovation are necessarily located in urban centers; however, as the above examples show, shifting emphasis onto the countryside in this way does little to alter the underlying socioeconomic or temporal frameworks used to understand it and in some cases even relegates parts of the countryside to the past. A more thoroughgoing interrogation of rural futures would require attending not only to the industrial timescapes of the ag-tech industry, but also the nonlinear ecological growth that it often subsumes.

What Koolhaas's remarks on the role of architecture do highlight, however, is a certain methodological reflexivity, whereby he employs a mode of futuring conscious of its performative effects. This is distinct from some of the speculative forms discussed in previous chapters, such as the predictive impulse underpinning geoengineering. For the most part, *Countryside, The Future* leverages effects that are reputational, in keeping with AMO's role as the research and branding arm of Koolhaas's larger operation, but insofar as it showcases systems and technologies that are actively being developed (some of them with AMO's involvement), it also creates expectations, sets shared agendas, attracts interest and potential investment, and, in the process, helps bring these possible futures closer to the present.

ENACTING POST-DEVELOPMENT

Elsewhere, it is possible to find a more critical meditation on the relationship between time, growth, and technological development, and a case study that moves further into the politics of enactment. Instead of a large-scale exhibition, the project took the form of a road trip between the two cities of Quito and Cuenca in Ecuador—a distance of 437 kilometers (272 miles), possible to cover in a single day by car. In this case, the car's fuel tank had been disconnected and strapped to the roof, stretching the journey out to a week, which was only possible with the aid of a variety of other automobiles, from lorries to RVs, whose drivers voluntarily towed the beleaguered vehicle across the country. The journey was the initiative of Adrián Balseca, an artist based in Quito whose fifteen-minute video *Medio Camino [Halfway]* (2014) is a curious reversal of the usual road movie tropes of frontiersmanship and self-sufficiency (figure 4.4), and part of a series of works, including *La Abundancia* (2014), pictured in the introduction, that explore relationships between transport, extraction, and time. The reversal in *Medio Camino* is quite literal in places, for example, when the truck is towed backward by slower vehicles, such as a tractor, and at one point a pair of horses. Balseca's own choice of vehicle is loaded with meaning as well: the Andino Miura 77, a small truck that was the first to be produced in Ecuador during the oil boom of the 1970s, and for many

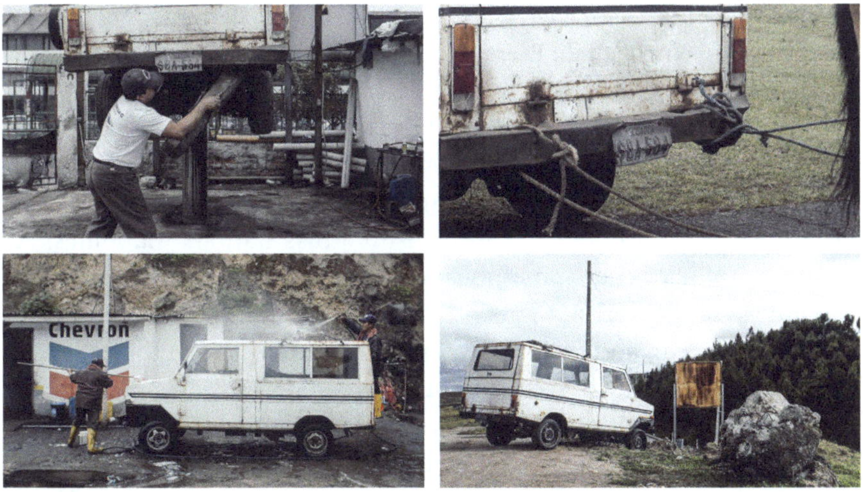

a symbol of modernity, industrialization, and economic development in the country. One of Balseca's talents as an artist is to expose the complex political significance of everyday objects, and in this film, the Andino does a lot of conceptual heavy lifting, being also the last vehicle produced under the military dictatorship of General Rodriguez Lara (1972–1976).

Yet it was under the reign of another of Ecuador's leaders that Balseca made this video work: president Rafael Correa (2007–2017). In many ways, the film stages a subterranean dialogue between these two periods in Ecuador's history, the first indexed to modernity, neoliberalism, and individualism, the second, to a complex renegotiation of this history that took place under the so-called "citizen revolution" led by Correa during his administration. One of the key terms that came to inform this renegotiation, and its eventual crystallization in the Ecuadorian constitution that Correa and others successfully took to the public in a referendum,

FIGURE 4.4

Adrián Balseca, video stills from *Medio Camino [Halfway]*, 2014, 15′41″ (loop). Courtesy of the artist.

was the concept of *buen vivir*, which is variously translated as "good living" or "wellbeing" in English. Buen vivir is itself a translation of the Quichua concept of *sumak kawsay*, rooted in an Indigenous worldview, although important differences exist between the two.[57] As de Zaldívar points out, the two terms occupy a complex discursive field, ranging from state-led utilitarian approaches that advocate the use of "buen vivir indicators," not dissimilar from attempts to measure gross national happiness, to vitalist ontologies that sometimes rest on essentialist claims of historical Andean and Amazonian peoples.[58] In each usage, emphasis is placed on communalism and reciprocity, principles embodied in the actions of those Balseca enlists in his cross-country road trip.

According to Eduardo Gudynas, another common feature of the increasingly divergent discourses on buen vivir is the challenge they pose to a singular conception of universal history.[59] In an article that helped bring the concept of sumak kawsay to greater prominence, the Indigenous-Amazonian intellectual Carlos Viteri Gualinga writes that the idea is resistant to Western notions "of a linear process of life that establishes earlier and later states of underdevelopment and development as a dichotomy through which people must move to attain well-being."[60] Indeed, critiques of buen vivir and sumak kawsay on the grounds that they are backward-looking or nostalgic have an unmistakably colonial overtone, positioning Indigenous-inspired worldviews as the past of a supposedly universal modernist future.[61] Building on this argument, Chuji and others argue that both sumak kawsay and buen vivir replace the narrow conception of linear time with a diversity of "multiple, parallel, non-linear and even circular, historical processes," with the addendum that "history" here refers not just to a human population, but to mixed socio-ecological communities.[62]

This embrace of temporal heterogeneity can also be felt in some of the literature on degrowth, which contributes to the same constellation of post-development imaginaries that have emerged in the last twenty years. There are numerous cross-pollinations between the two, with widely circulated reference books such as *Degrowth: A Vocabulary* and *Pluriverse: A Post-Development Dictionary* seeking to build conceptual bridges that emphasize commonality without erasing differences.[63] For many advocates of buen vivir, degrowth might appear narrowly economic in its focus, and

Eurocentric in its orientation.[64] Nevertheless, in both movements, there is a shared concern with how linear temporalities of development might be replaced with alternative configurations of pasts–presents–futures that are placed in the service of social and environmental justice. The two movements also overlap in their means of achieving such political ends, which embody a politics of enactment that views futures as realms of possibility that can be brought into conversation with pasts and presents, rather than as potential threats to hegemonic order to be predicted or pre-empted in advance.[65]

In the case of buen vivir, enactment can be seen to function in two ways. One of the more common uses of the verb "to enact" occurs in the context of changes made to a national or international legislative framework. When a law or parliamentary bill is enacted, it takes on a new reality, imbued with the power to effect change through established legal routes. In Ecuador and Bolivia, a certain conception of buen vivir has become enshrined in the constitution, with protections for "the rights of nature" in particular drawing international attention. Although these principles have been slow to be translated into legal cases—a point illustrated in the extractivist agenda subsequently pursued by Rafael Correa, in direct contradiction to principles of buen vivir—weaving them into the constitutional fabric of a nation-state is more than a symbolic gesture; it is an attempt to cultivate environmental futures within the heart of government.[66]

In a second sense, both buen vivir and degrowth seek to enact socially and environmentally just futures through aligning political ambitions with the practical means of achieving them. This means–ends alignment has, in other contexts, been conceptualized as "prefigurative politics," which Jeffrey and Dyson define as "the self-conscious channelling of energy into modelling the forms of action that are sought to be generalised in the future in circumstances characterised by power, hierarchy, and conflict."[67] As the authors note, from the perspective of the Global North, the idea of modeling forms of political action in this way has its origins in anarchist politics of the late nineteenth and early twentieth centuries, as well as feminist scholarship and activism from the 1960s to the present.[68] Frequently cited examples of prefigurative politics include alternative education initiatives and the Occupy movement, although scholars have also used the term in discussions of both degrowth and buen vivir.[69] Notably, it can also be discerned in the "emancipatory" history of futures studies (albeit not always in name), where figures such as Johan Galtung sought to counter elitism and "avoid traditionalism" in the constitution of the WFSF in the 1970s by embedding principles of democracy, openness, and plurality in the way the organization functioned. The concept also continues to have relevance to futures studies today, often in discussions of how activist groups or social movements embody the values they want to see realized outside of their organizations and networks.[70]

While both prefiguration and enactment involve ways of actively and self-consciously bringing futures into the present, what makes enactment a useful analytic here is its links to performance and improvisation, and its relationship to questions of scale and scalability. On the first point, prefigurative politics is often conceptualized as the re-creation of a blueprint of a desired future in miniature form, whereas enactment implies a relationship to futures that is less scripted and provides space for ideas to be progressively worked through—even in the above case of constitutional reform, where every case sets legal precedents for the next.[71] On the question of scale, as Jeffrey and Dyson's definition indicates, a key characteristic is the replication of a model at various magnitudes, following the logic of a synecdoche, in which a part comes to stand in for the (future) whole.[72] Prefigurative politics models social arrangements that are intended to be generalized, expanded, and scaled up, beginning with "a miniature version of the society desired," whereas enactment engages with relationships of scale that are less direct.[73]

This much is already clear from *Medio Camino*, which uses a humble gesture to enact principles of buen vivir without intending for the action itself to be scaled up. In its references to moments in the political history of Ecuador, communicated through physical objects, it improvises with historical materials, exploring a space somewhere between symbolic gestures and concrete actions. By taking gas out of the tank of the Andino, Balseca signals toward a post-extractivist future while simultaneously enacting a system of solidarity economics that would serve as its desirable social accompaniment. While the political effects of such gestures might be considered limited, it succeeds in collapsing the distinction between representing a future that has yet to arrive and carrying out concrete actions that embody some of its ideals, without for that matter seeking to create a world-in-a-grain-of-sand that captures its totality.[74] In contrast to *Countryside, The Future*, Balseca does not enact by means of setting in motion expectations for technological innovation, but rather by producing an aesthetic intervention that combines thought and action around a structure deliberately at odds with the linear temporalities of development.

FOR A GROWTH THAT DOES NOT SCALE

The emphasis on non-monetary transactions in some of Balseca's work is amplified in the practical enactments of many artists outside of Latin America, too. One of these is Kathrin Böhm, who for the past few decades has worked collaboratively as part of various ongoing groups including Myvillages (co-founded with Wapke Feenstra and Antje Schiffer in 2003), the Rural School of Economics (also established with Feenstra, in 2021), the art and architecture collective Public Works (founded with Sandra Denicke-Polcher and Stefan Saffer in 2004), and the Centre for Plausible Economies (initiated with Kuba Szreder in 2018). On the occasion of an exhibition by Böhm in 2021 at London's The Showroom gallery titled *Compost: For Future Use*, the artist took the opportunity to sift through some of the material and conceptual residue from these projects, reintroducing it into economic circuits of exchange through diverse means.

To take one example: In a corner of the gallery, a large tractor tire is propped up against the wall. Böhm explains that there are two possibilities for the future of the object: Either it is sold for around £7,000 to a museum or collection such as Arts Council England, or it is used as non-art and moved to a local adventure playground for free (figures 4.5 and 4.6).[75] Elsewhere, it has functioned as both gallery seating and symbolic shorthand for the system of technocentric monoculture farming that much of Böhm and Myvillages' work criticizes. This object, like many others piled, stacked, and arranged in the room, encapsulates the artist's attitude toward the use value of art, which she characterizes as "a weird tool that people make use of in various walks of life."[76] There are plenty of other objects on display which could be used as everyday items and/

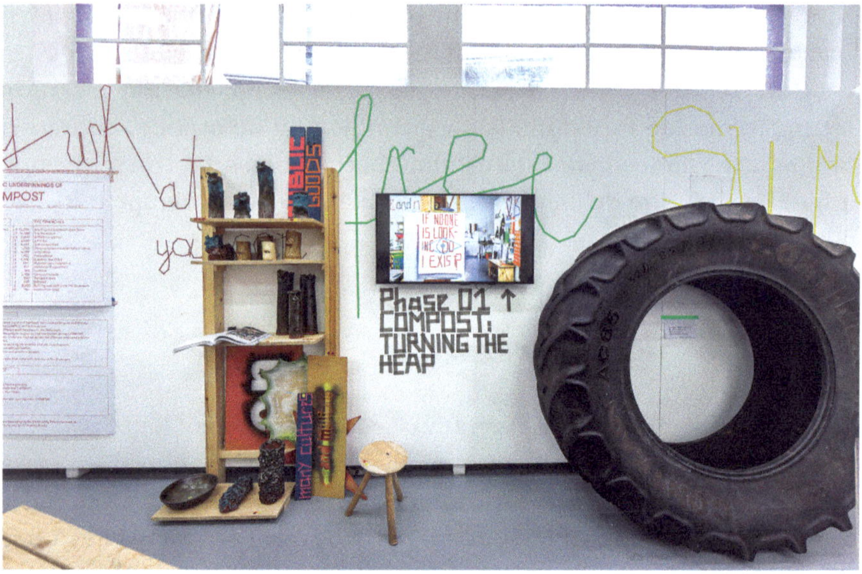
free
Phase 01
COMPOST:
TURNING THE
HEAP

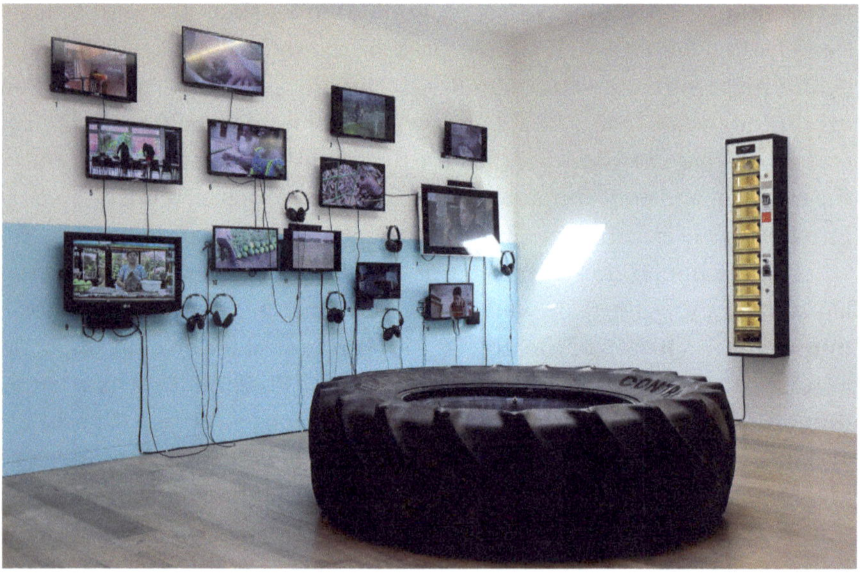

or artworks, including ceramics, printed matter, and items of furniture made from the detritus of her past installations. Some of these are for sale; some, available to take away for free; and some, available for barter. As its title signals, the exhibition is built around the structuring metaphor of compost making, of "turning the heap" so that cycles of creative renewal might be facilitated by engaging with what might otherwise have been discarded.[77] Items are piled up in such a way that discrete borders between them dissolve and component parts become available for new use. On the odd occasions that traditional modes of spectatorship are encouraged by the artist—as they were in an exhibition featuring Myvillages in Bratislava that ran concurrently with the London show—it is often with another goal in mind.[78] In this respect, Böhm is transparent about her intentions for holding the London exhibition, which is not especially representative of the broad range of socially engaged activities that constitute her individual and collaborative artistic practice: It is an experiment with one of the functions of a gallery exhibition, to constitute and increase the market value of objects exhibited as art, the proceeds of which will go back into some of the collective projects where the objects originated, such as the cultural enterprise "Company Drinks."

Set up in 2014 as an art commission seed-funded by the Create Art Award, and later registered as a social enterprise, Company Drinks is a long-term piece of social infrastructure considering every stage of the production, distribution, and consumption of a simple and accessible product: bottled drinks (figure 4.7). Reconnecting to the social history of "going picking" (also known as "hopping"), a practice in which

FIGURE 4.5

Kathrin Böhm, *Compost: For Future Use*, The Showroom, London, 2021. Photo by Dan Weill. Courtesy of the artist.

FIGURE 4.6

Myvillages, *Setting the Table: Village Politics*, installation at Whitechapel Gallery, London, 2019. Photo by Wapke Feenstra. Courtesy of the artist.

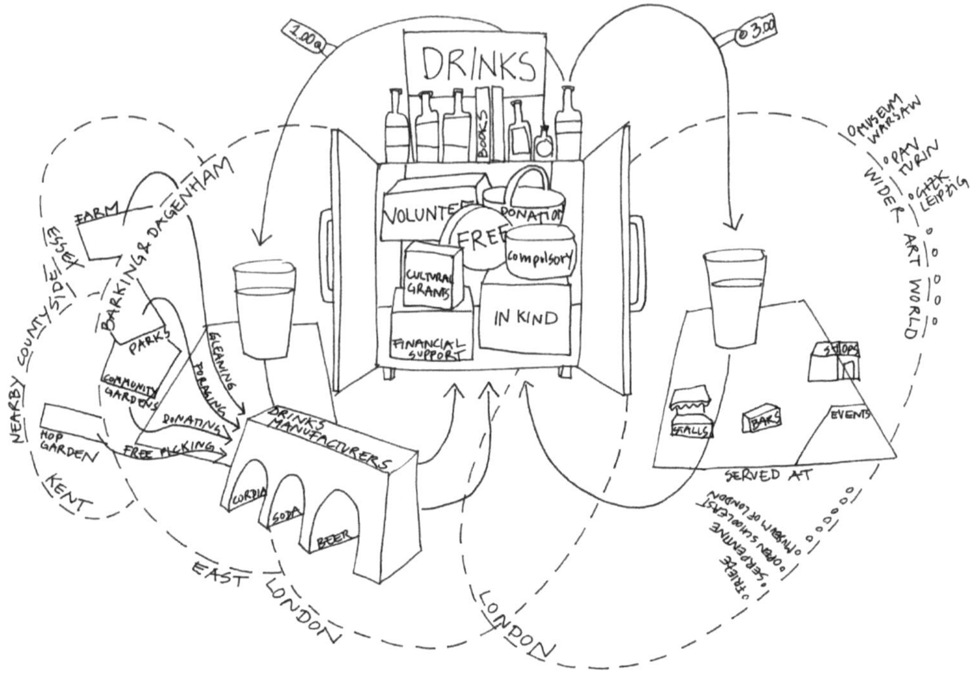

working-class families from London's East End would form a temporary workforce each summer to carry out the annual hop harvest in Kent, Company Drinks seeks to establish new economic and social relays between London and its neighboring rural environments. At the height of hop production in Kent between the middle of the eighteenth and nineteenth centuries, a large proportion of the county's land was taken up by the crop, with as many as 200,000 temporary workers arriving from outside the county, a large proportion on specially provided trains from London. As much as "going picking" was primarily a way of supplementing income, it

FIGURE 4.7

Kathrin Böhm, *Company Drinks Business Model Diagram*, 2018. Courtesy of the artist.

was also accompanied by a distinctive sociality, with women and children making up the bulk of the workforce.[79] Many of them would use it as an excuse to take their children off to the countryside for a working holiday, where families lived in makeshift hoppers huts they would return to year after year.

Whilst operating on a much smaller scale, and not primarily geared toward the production of beer, Company Drinks seeks to build a similar sociality around how economic relations of production and consumption are enacted. As one promotional video puts it, "The production cycle becomes like a new community space."[80] Trips are made to forage, pick, and "glean" drink ingredients from fields and hedgerows. Workshops are held with children from local schools in Dagenham on drinks mixing, branding, and distribution. And when the product does finally make its way to market, consideration is given to the communities it encounters. In Dagenham, one of the most deprived boroughs of London, the drinks sell for one pound per bottle. When Company Drinks set up a stall at the annual Frieze art fair in London, they sell the drinks for three times the price.

In its historical reference to the practice of hop picking, as well as its embrace of pre-capitalist modes of exchange such as barter and commoning, Company Drinks would appear to be a somewhat nostalgic initiative, reaching back to a time when labor practices facilitated a strong sense of class and community identity. To recall the language of the photomontage in *Countryside, The Future*, this is a version of agrarian existence in which "everyone—man, woman, child—knows their place." In both representations, agricultural work was primarily manual rather than mechanized too, as it is now on all but one of the remaining hop farms in Kent. And yet, the project is far from oblivious to contemporary capitalist labor practices. In a film component of the project, *Foreign Pickers*, seasonal workers from eastern Europe were brought into contact with some of the east Londoners who took part in annual hop harvests, their conversations, anecdotes, and exchanges emphasizing cross-cultural and cross-generational links based on experience rather than national identity (figure 4.8). Created in 2016, a week after the UK's referendum on membership of the EU, the film also subtly invites contemplation on the futures of agriculture in the UK if labor regulations were to change—an eventuality that continues to cause problems in the country many years

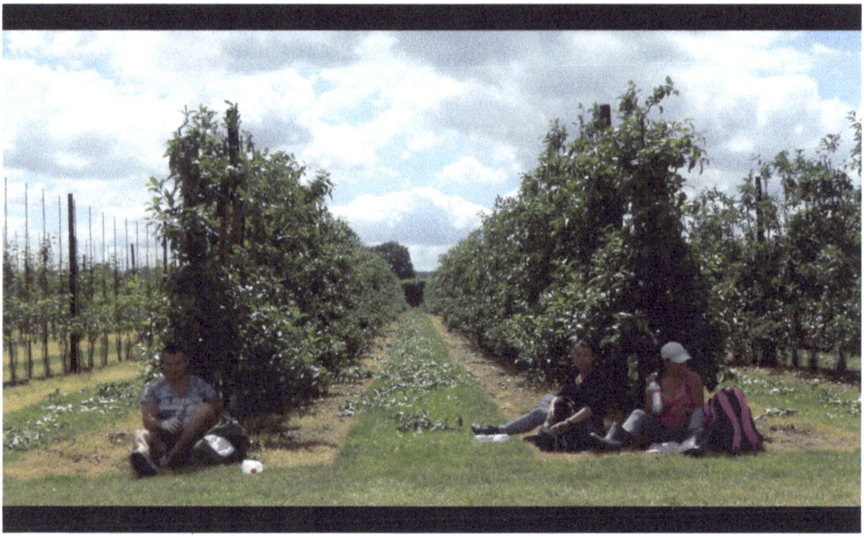

after Brexit. In reactivating the history of hop picking and connecting it with the uncertain futures of migrant labor today, Böhm refuses the kind of modernist teleology embedded in *Countryside, The Future*, in which the past is treated as a static backdrop. Instead, the project facilitates a subtle form of time-travel, facilitating seasonal worker solidarity across borders in the process. This is also, decidedly, not a body of work "about" the countryside, or climate change in any direct sense. Individually and collectively as part of Myvillages—a group that has mounted as many exhibitions in locations such as barns, village halls, and market stalls as it has in art galleries in urban centres—the projects use a diversity of approaches to critique what Myvillages calls the "urban gaze," intervening in

FIGURE 4.8

Myvillages and Company Drinks, still from *Foreign Pickers*, 2016. Concept by Kathrin Böhm, production and edit by Sue Giovanni. Commissioned by Delfina Foundation/ Politics of Food. Courtesy of the artist.

the infrastructural linkages between the urban and the rural, rather than packaging the latter up as an artistic product for predominantly urban cultural consumption.

In a number of the pamphlets accompanying her projects, Böhm makes explicit reference to the feminist economic geography of J. K. Gibson-Graham, the joint authorial project of Julie Graham and Katherine Gibson. Their "diverse economies" perspective privileges the kinds of non-market-based transactions that take place alongside and in tension with more traditional forms of capitalism. Socioeconomic forms such as reciprocal exchange, gift economies, and worker cooperatives are often figured as the leftovers of outmoded economic systems, but for Gibson-Graham, these forms are the starting point from which to enact post-capitalist futures, being "exactly what will sustain us in times of financial and ecological crisis."[81]

Gibson-Graham has been criticized for being overly pluralistic in their account of non-capitalist social relations, as if simply providing an inventory of diverse economies were in itself a radical act. Furthermore, as a long history of feminist political economic thought and activism lays bare, some modes of exchange are not desirable simply by virtue of taking place outside of the formal economy. Their openness to multiplicity is, however, a deliberate attempt to counter the "overdetermination" of capitalism as the socioeconomic lens through which everything else is viewed.[82] As Gibson-Graham puts it in *A Postcapitalist Politics*, "Our language politics is aimed at fostering conditions under which images and enactments of economic diversity (including noncapitalism) might stop circulating around capitalism, stop being evaluated with respect to capitalism, and stop being seen as deviant or exotic or eccentric—departures from the norm."[83] Building on discussions in previous chapters, it is possible to recognize an element of performativity at play here, mobilized as an attempt to bolster the power of non-capitalist social relations by shifting the framing devices used to construct them as material-discursive realities.

The same argument could be applied to the pragmatics of how economic growth is spoken about. As implied by Dale's concept of "the growth paradigm," GDP is more than simply an economic indicator used at the national scale. Its discursive power penetrates deep into the social body, absorbing practices that might not otherwise be pulled into its wake.

Artistic output is far from immune to growth-based thinking either, with careers made or unmade on the ability of artists to navigate a fast-moving reputational economy premised on increasingly ambitious projects, with art's positive political and environmental impact often overstated. In some ways, Böhm's exhibition at The Showroom represents a withdrawal from the art world's growth paradigm, in an effort to slow down, take stock, and nurture existing projects rather than ceaselessly move onto the next. And yet, the exhibition's very existence is testament to a canny attempt to use growth against itself, to effectively "buy time" by raising the market value of her work and redistributing the profits. In this way *Compost: For Future Use* is an example of a diverse economies approach being enacted through projects that pluralize the modes of exchange usually accommodated within a gallery setting. More than a set of blueprints for how the political economy of the art world might be improved, it takes a series of small practical steps to make post-capitalist futures a reality, using artistic competences to weave relational webs between the rural and the urban, art and non-art, pasts and futures.

A key feature of Böhm's work that marks it out as enactive rather than prefigurative again hinges on the concepts of scale and scalability. As she puts it in a short illustrated text titled "The Penfold Principles," co-written in public with Kuba Szreder and inspired by the cooperative movement's "Rochdale Principles" of the mid-nineteenth century, "The ambition of [interdependent] art worlds is to create self-defensive structures against . . . extraction, aimed at ubiquity and equity—not to prop up careers, or to scale up organisations, but to spread out, sustain and enable."[84] Spreading out and scaling up here represent two divergent paths: the former based upon sharing resources and creating lateral connections (across both space(s) and time(s)); the latter, on presumptions of scale-invariance according to which the characteristics of an organizational structure or model can be replicated at different sizes with little modification.[85] Spreading out involves the nonlinear linking of spatiotemporal particularities, for example, connecting migrant fruit pickers in the UK with the working-class labor force of the nineteenth and early twentieth centuries. Scaling up creates linear relationships between the past and present, which serve as a backdrop against which projections and forecasts of future growth can be plotted. Critiques of political projects that resist pressures to scale up are sometimes viewed as "stuck in the local" or tending toward a politics of social enclosure within "premium ecological enclaves."[86] However, projects such as Böhm's demonstrate that self-defensive structures can be enacted in ways that connect particularities through webs of interdependence. These particularities are not ring-fenced visions of desirable societies in the miniature but pockets of experimentation that gain power through connection. In doing so, they create what I have called "patchwork futures," providing examples of the generative multiplicity of practices that resist a unifying image of the future. To demonstrate this point, it is worth briefly surveying a number of projects that share this enactive quality, recovering agrarian pasts and helping to establish trans-local linkages between organizations

The Spanish group Inland, for example, initiated by the artist Fernando García-Dory, explores neo-pastoralism through activities that have grown out of an annual program to combat a decline in the number of shepherds working in northern Spain. As well as organizing educational activities, the project has spawned exhibitions, books, a "world gathering of nomadic and transhuman pastoralists," and technological innovations such as a "flock protection system" intended as a "creative solution to the age-old pastoral rivalry of the shepherd and the wolf."[87] Today, Inland produces artworks, cheese, and other agricultural produce, and organizes study groups and exchanges from its two bases in Madrid and a hamlet it has helped recover in northern Spain. As well as fostering linkages and interdependencies between rural locations within Spain, it participates in international networks such as the Confederacy of Villages, building upon earlier works such as García-Dory's *The Cabrach Chart* (2013), a project that linked the politics of land use in Cabrach, Scotland, with a radical landless farm workers union in Andalusia, Spain. The aim was to embody Paulo Freire's methodology of "peasant to peasant" exchange in nonhierarchical situations—not dissimilar to the kinds of exchanges facilitated by agricultural organizations such as Via Campesina—and resulted in a series of speculative proposals for land reform.

In the Occupied Palestinian Territories (OPT), another project at the intersection of art, agroecology, and the politics of territory has worked to recover agrarian histories in the service of emancipatory futures. Sakiya, an initiative of the artists Sahar Qawasmi and Nida Sinnokrot, is situated on a hillside in the village of Ein Qiniya, northwest of Ramallah. In an area of the West Bank heavily controlled by Israel, which imposes restrictions on the movement of heavy machinery, the kind of scalable agricultural techno-futures imagined in Rem Koolhaas's exhibition are foreclosed. Some of the most productive land lies behind barriers and checkpoints in the Jordan Valley, where it attracts large subsidies from the Israeli government and is irrigated by the Jordan river.[88] In a familiar story of settler-colonial oppression, which pre-dates the genocide in Gaza, Palestinian farmers, whose access to the river has been cut off, enjoy no such subsidies from the Palestinian authority, and consequently, the market has been flooded with cheap produce from settler communities. This crippling dependency, exacerbated by climate change, makes it difficult to persuade Palestinians to get into farming on the basis of economic motives alone. Connecting farmers, local villagers, visiting artists, students, writers, and craftworkers, Sakiya enacts an alternative value system around the land in the face of these challenging circumstances. The aim is not to scale in the manner of intensive farming but to cultivate experimental knowledge grafted onto Palestinian agrarian traditions that can spread, attuned to the changing socio-environmental conditions of the land.[89]

While both of these projects have international reach, perhaps the most high-profile exploration of enactive politics in recent years was the fifteenth edition of the Documenta exhibition, held every five years and generously funded by the German state—a funding arrangement that has contributed to political controversies and censorship.[90] The 2022 edition was curated by the artist group ruangrupa (spelled with a lower case "r") around the central organizing principle of *lumbung*, an Indonesian word for a collectively managed rice barn. True to the collective spirit of lumbung, ruangrupa invited fourteen core collectives to take part in the exhibition, who in turn extended invitations to others, the majority of whom were located in the Global South. In contrast to the usual practice of commissioning artists to produce new work for the exhibition, and inspired by ideas of a basic income for artists, ruangrupa instead paid each of the original fourteen core members 25,000 euros (26,300 US dollars) upfront in recognition of the work they had already done in their communities, as well as a production budget of 180,000 euros per group; the remaining participating artists received 5,000 euros (for individuals) or 10,000 euros (for collectives) upfront, in addition to 60,000 euros for production. Each participant then put 20,000 euros of their budget into collective ownership, governed by democratically made decisions, and a working group was established to facilitate experimentation with alternative economic structures resulting in "lumbung Kios."[91] This network of kiosks and partner businesses diversified income beyond state funding, allowing for non-monetary exchanges to trade affordable objects, the profits from which were divided between long-term projects nominated through an assembly forum, or *majelis*.[92] Many of the participating groups themselves experimented with alternative economic structures, for example, the Palestine-based group Question of Funding, whose *Dayra* model aims to support "deep locality" by establishing interconnected circles of trust between cultural producers and small businesses using a non-monetary model that encourages circulation rather than accumulation.[93] As this sketch of the exhibition suggests, post-capitalist economics were enacted at various organizational levels of Documenta 15, and, crucially, this was done not through scaling one model up but by encouraging collectivism without homogeneity, creating a patchwork of diverse economies held together by shared values rather than mediated through market mechanisms and their attendant myths of self-regulation.

What this brief survey of artistic projects reveals is a shared politics of enactment, wherein entwined futures of food and cultural production are lived as realities in ways that seek connection without homogeneity. In cases such as Company Drinks, Inland, Sakiya, and many of the projects participating in Documenta 15, connections are established in time as well as space, diverging from the kinds of future-oriented practices that position the past as a stable ground from which growth and technological innovation can be engineered through successive economies of scale. While few of the projects are labeled with the word "degrowth" explicitly, they each embody the chronopolitical drive of a movement that is as adept at moving between pasts–presents–futures as it is at accommodating different scales of operation. When considering the aims of this book more broadly, this work has the potential to complement discussions of prefiguration within futures studies by introducing important considerations of scale and connection that diversify existing pathways to environmental futures, embodying a "patchiness" better aligned to the dynamics of ecological growth.

NEITHER BACKWARD NOR FORWARD

From its early life as an activist slogan in the 1970s to its current function as the watchword of a developed political philosophy, the word "degrowth" has provided a focal point for various initiatives that take climate and biodiversity crises as their point of departure. Whilst the primary focus of degrowth is undeniably economic, this chapter has shown that a more diffuse culture of degrowth is also gaining visibility, forged from unlikely alliances between activists, agroecologists, and artists, comprising a loose set of projects, organizations, and initiatives that enact the possibilities for life outside of the linear growth paradigm. These diverse actions cannot be flattened into one perfectly coherent position, nor can they be traced back to a single point of intellectual origin, as demonstrated by the complementary and yet crucially distinct traditions of degrowth and buen vivir. As Gibson-Graham's diverse economies approach also advocates, this pluralism is an important feature of post-growth imaginaries, constituting a "world of many worlds" that contains within it diverse speculations for futures beyond extractivism.[94] This too is a speculative infrastructure of sorts, albeit one that takes a radically different form from the infrastructure that enshrined the doctrine of economic growth in the middle of the twentieth century. Here coherence is not established through institutions and initiatives that "harmonize" economic policies and generalize industrial time, but in and through encounters between historical and contemporary practices that sustain socio-ecological flourishing.

In contrast to transnational corporate and state-led organizational infrastructures that use anticipatory politics to pre-empt threats to hegemonic order, cultures of degrowth and post-development harbor within them a mode of futuring that could be characterized by enactment.[95] Like the other modes of speculation discussed in this book, such as prospecting or scenario planning, enacting involves building cultural relationships to futures. In all cases, this relationship is inflected by a performative dimension whereby a prediction, plan, or enactment intervenes in the unfolding of the present, helping to constitute futures as objects of experience in the process. What sets enacting apart from prediction and scenario planning is that this performative dimension is embraced rather than disavowed, something that connects it to the "emancipatory" tradition of futures studies initiated in the '60s and '70s. To enact is not to produce a dispassionate prognosis on what *will* happen in the future, as if the speculation had no effect on the course of events. Nor is it to make cultural objects *about* futures. It is to actualize imagined futures in the present, to create a "living laboratory" for tools and techniques that embody the futures their creators desire.

Perhaps by virtue of their proximity to the present, many of the diverse post-growth imaginaries discussed in this chapter do not appear futuristic in the ways many have come to expect. Shaped by the techno-imaginary of mainstream science fiction, it has become a habit of mind to populate futures with increasingly sophisticated technologies, while the underlying socioeconomic structures remain curiously unaltered. As Ursula K. Le Guin wrote in the early 1980s, "It seems that the utopian imagination is trapped, like capitalism and industrialism and the human population, in a one-way future consisting only of growth."[96] Many of the artworks discussed in this chapter represent a conscious break with this one-way future. To frame the post-growth imaginaries they enact as nostalgic or backward-looking, simply because they work with existing tools and technologies, is to fall back into a modernist telos of progressive technological innovation. In rejecting the false choices between backward and forward, localized folk politics or planetary solutionism, it becomes possible to build a speculative politics that establishes new relays between pasts, presents and futures, enacting alternative worlds that take responsibility for the futures they bring into being.

CONCLUSION

A LOCKPICKER'S GUIDE TO CLIMATE FUTURES

"It's not climate change—it's everything change," wrote Margaret Atwood in an article in 2021, echoing the many other writers and theorists who have remarked on the disorientating quality of the first decades of the twenty-first century and the scale of response required.[1] At present, the degree to which "everything change" is experienced falls unevenly across the world, with many lives in the Global South already being destroyed by environmental catastrophes, and more affluent nations remaining largely cushioned from the worst effects—a cushioning that, in turn, ensures the continuation of industries and lifestyles that exacerbate anthropogenic factors. Even more so than during the rise of futures studies in the postwar period, today's era is one in which a "bonfire of certainties" means that speculative tools that rely on the past as a guide to the future lose their predictive edge. Amid this uncertainty, a pragmatics of foresight emerges that focuses on specific corporate or policy gains, often mobilizing discourses of crisis and emergency to extend unequal power relations across temporal horizons. As Ben Anderson puts it, this is an anticipatory politics that works by "(re) making life tensed on the verge of catastrophe in ways that protect, save, and care for certain valued lives, and damage, destroy, and abandon other lives."[2] While certain nation-states and organizations have some ability to modulate sequences of events as they unfold, others fall on the wrong side of this strategic divide, experiencing the full volatility of climate catastrophe and biodiversity loss with little means of protection.

Yet, in this landscape of perpetual change and cascading uncertainties, there are many social and material processes that nevertheless feel stubbornly entrenched. Large-scale transitions from one energy regime to another proceed in fits and starts, impeded by the short-term thinking of politicians and pressure from powerful lobbies and reactionary movements. At a smaller scale, behavioral norms and persistent habits make it difficult for individuals and social groups to imagine anything outside of business-as-usual existence. The apparent intransigence of some processes in the face of wide-scale socio-environmental change has been described at various points in this book, sometimes using the terms "lock-in" or "path dependency."[3] There are numerous examples of lock-in at work in the preceding chapters—some with an explicit socio-technical grounding, others more behavioral or discursive—and casting a backward gaze will help pull some of its threads together around a concept that has the potential to distill the central argument into a more programmatic form.

Thinking from the middle outwards, many of the engineering practices discussed in chapter 3 provide particularly clear examples of lock-in, often in ways that mix the material and discursive. In the case of Mathur and da Cunha's examination of hydraulic engineering in Mumbai, the borders that separate land and water are shown to be the product of both the hard materiality of drainage infrastructure and deep-rooted conventions in viewing the landscape, for example, through colonial cartographic practices. Material-discursive path dependencies are also evident in practices more closely associated with geoengineering. The history of attempts to modify the climate are shown to be surprisingly long, with many "new" technological solutions to climate change, such as stratospheric aerosol injections and ocean fertilization, rooted in Cold War era military and scientific endeavors. The fact that these ideas have hung around for so long is not testament to their unique power or efficacy, since many techniques of geoengineering remain untested; rather, their longevity can be attributed to their rootedness in enduring knowledge infrastructures and epistemic cultures, such as that of a conception of the world as a giant system of inputs, outputs, and feedback loops, discernible by adopting a macroscopic view.

The geoengineering technique of stratospheric aerosol injections provides a particularly stark example of path dependency and its capacity to lock societies into certain futures and exclude others. While this method was first deployed in the 1960s and '70s for the purposes of cloud seeding, it has yet to be deployed at scale for the purpose of managing solar radiation. If this takes place as some of its proponents intend, future generations could be forced to continue pumping aerosols into the atmosphere for fear of "termination shock," in which the abrupt cessation of such activities would likely result in climate impacts expected to unfold over a period of decades, ravaging the planet in a matter of years, with apocalyptic ramifications for humans and biodiversity.[4] The environmental futures that result from this techno-solutionist proposal splinter into two, so that a false choice is presented between a burning world that results from doing nothing and dependency on planetary weather modification in the medium to long term. Catriona McKinnon explores the idea that current research on geoengineering may already be generating what she calls "cognitive" lock-in through the way it frames techno-fixes as viable responses to climate change. This not only diverts funding away from strategies associated with rapid decarbonization and sidelines socioeconomic proposals for reducing the material throughput of the economy but also creates a "new normal" in which such ideas are progressively seen as the only remaining option.[5]

The idea of cognitive lock-in helps reinforce the idea that visual cultures can provide fertile ground for the growth of many of the technological infrastructures upon which societies have come to rely. In chapter 3, art and visual culture are shown to both exemplify some of the visual logics that support an engineering mindset—as is the case with Peter Fend—and provide a parodic counterpoint that uses the macroscopic view against itself. Such examples embody one of the key roles accorded to art and visual culture in this book: its capacity to isolate entrenched processes and dynamics and offer these up as objects of reflection. For while those trying to change any given socio-technical object or system may be all too aware of its intransigence, often, the roots of such systems reach deeper than may at first be apparent. By bringing these roots to the representational surface, not only can artworks reveal overlooked historical processes, but they can also show how these processes continue to exert influence on how environmental futures unfold in the present.

Many of the creative practices discussed in chapter 2 also embody this argument, revealing path-dependent processes and visual regimes than might otherwise hide in plain sight. Here, lock-in is seen to operate in the prospecting activities of companies that sink so much capital into finding and exploiting resources that they subsequently become difficult to abandon. Even before capital congeals into physical infrastructure, prospecting works performatively on environmental futures by creating a series of expectations that help broker relationships, set shared agendas, and attract investment. Such activities are often themselves layered upon colonial histories and entrenched asymmetries involving additional lock-in dynamics of debt and dispossession. Visual cultures are shown to be central to this process, framing the world as a basket of resources, a collection of specimens, or, in the case of Timothy O'Sullivan's survey photographs, a depopulated frontier awaiting exploitation. Against this visual culture of prospecting, a visual culture *about* prospecting that includes the work of On-Trade-Off and Edith Morales shows that a "view on the view" could unpick the visual locks of extraction and weaken the simplistic equation of verticality with domination.

The configuration of images, actors, and processes in chapters 2 and 3 demonstrates how constraints are placed on environmental futures, sometimes without any direct intent. The same is true of the speculative infrastructure discussed in chapter 1. While scenario planning is shown to have crossed numerous ideological and geographical divides since its inception in the 1950s, when entrenched within corporate, governmental, and scientific organizations, it loses some of its imaginative power. This has had the effect of narrowing the discursive range in which environmental futures can be imagined to those that are "policy-relevant" or most likely, in line with what Ahlqvist and Rhisiart have framed as the "utilitarian" trajectory within futures studies.[6] For Eshun, this utilitarian side is also bound up in the dogma of modernization that locks nation-states onto developmental paths to futures defined for them by countries in the Global North. Projects by the likes of Arata Isozaki, Peter Watkins, Helen and Newton Harrison, Black Quantum Futurism, and others are positioned as a powerful antidote to institutional and ideological inflexibility. While Isozaki and Watkins are rooted in the same historical context as futures researchers with explicitly progressive motivations, many of whom contributed to organizations such as the WFSF, other creative practitioners are shown to work within an "expanded field" of scenario thinking. The idea of an expanded field can be understood in two senses here: 1) as a call to look beyond the narrow institutional history of scenario planning and consider how this history is entangled with a wider creative imaginary of multiple futures (whether by historical association—as in the case of Isozaki—or through epistemological affinities); and 2) as a demonstration that creative practice can broaden the parameters of what is considered a valid form of future-making. This second point is both fundamental to the ambitions of this book and implicates a much wider field of creative practices that can transform futures from objects of rational planning into experiences of aesthetic multiplicity. The multiple futures of scenario planning are shown to extend into multiple pasts as well, with certain currents within Afrofuturism seeking to dislodge the idea of time as a reliable, unbroken line between past–present–future, with the manifold cognitive, institutional, and socio-technical path dependencies this entails.

Undergirding many of the mainstream environmental futures in chapters 1–3 is an assumed backdrop of continual capitalist innovation and economic expansion. This backdrop comes into the foreground in the final chapter, which explores attempts to actualize—or "enact"—futures in the present that depart from an institutionally enshrined paradigm of perpetual economic growth. The hegemony of the growth mindset is discussed in terms of its reliance on the emergence of an abstracted industrial time, and exemplified in the exhibition *Countryside, The Future*, which is shown to embody a growth mindset in how it produced expectations of techno-capitalist projects with performative effects. By contrast, selected works by Adrian Balseca, Kathrin Böhm, and others establish new relays between pasts, presents, and futures, and explore alternative economic practices through practical enactment. It could be argued here that such alternative economic practices can themselves fall prey to lock-in; however, by embodying an improvisatory politics of enactment rather than creating blueprints for desired futures in miniature through prefiguration, a degree of flexibility is often built into these projects from the beginning. While none of the artists discussed in this chapter have explicit links to futures studies as a defined field of study or expertise, they nevertheless provide inspiration for how critical futures work might embrace its performative power and explicitly acknowledge embedded cultural and political commitments.

As the varied examples of lock-in and path dependent processes spread throughout this book demonstrate, all manner of ideological structures and linguistic norms can become self-reinforcing.[7] From an individual's consumption habits to transnational trade agreements that create favorable regulatory environments for extraction, processes of entrapment can seem like a ubiquitous feature of contemporary life. One of the central arguments of this book is that the bridges and pathways we build to futures can themselves fall prey to lock-in processes. Paying attention to very different contexts of creative and intellectual production, all four chapters of *Futures in the Making* share the aim of understanding where patterns of thinking about environmental futures get stuck. The concept of lock-in gives a name to these processes, but it also points to the ways in which alternative modes of thinking are actively locked *out*, with communities of living beings "defutured" in the process. The various examples in the preceding pages reveal the subtle ways in which power can percolate through infrastructure in a way that both includes and excludes. Self-reinforcing processes push certain activities into the foreground and let others fall into the shadows, where they enter terminal decline. One of the reasons that changes to behavior in areas such as energy infrastructure or transport are so incremental is because path-dependent processes work to discourage radical change, not by decree but through what Foucault memorably called the "capillary action" of power through the social body. Lock-in processes take root in the smallest pockets of life, determining the conditions of possibility for activities that extend far beyond their apparent sphere of influence, so that the task of creating a new path becomes a matter of both practical and imaginative difficulty.

What I call "speculative infrastructures" in this book are comprised of tools and techniques that act upon environmental futures, shaping and scripting the way they are constituted as objects of the imagination, and providing points of orientation for a broader set of environmental activities. The more powerful they become, the more ordinary such infrastructures seem to be, to the extent that it is now difficult to imagine a world in which economies do not rely upon the creation of resource frontiers, or businesses, governments, and climate researchers do not filter futures through the lens of scenario planning. And this unobtrusiveness of many speculative infrastructures is precisely what allows us to forget that they are at work creating boundaries between the possible and the impossible that can become increasingly brittle over time. This boundary work has the effect of closing the imagination to certain possibilities, anchoring it to the drag weight of a diminished past, and constricting the discursive space in which possible futures can be imagined.[8] Whether concerned with the blunt materiality of mining infrastructure or the infrastructures of feeling that shape scenario planning, an urgent question becomes how to prevent the tools used to imagine possible futures from delivering the same pre-packaged results.

This book uses art and visual culture to answer this question and, in these final pages, positions it amid a landscape of ubiquitous social, cultural, and technological lock-in. As the assorted creative projects show, both critical and propositional modes of practice are required. In its critical capacity, art can bring the political disposition of techniques of futuring into sharper relief, acting like a dye in the capillaries of infrastructural power. In line with much critical social science work (CFS and STS included), this diagnostic activity can reveal how ostensibly technocratic processes contain unspoken assumptions and embedded values. In doing so, it can recover forgotten histories, looking to the past to understand the interconnected infrastructures that enmire the present. It can also cast a contemporary gaze at infrastructures still in formation to identify the "critical junctures" where processes of entrapment can be avoided. Early proponents of futures studies such as Johan Galtung were already aware of this fact, calling for futures researchers to act as a "defensorate for future generations" and "watch out for developments that may enslave future generations."[9] The same role could be accorded to creative practitioners. Given the importance of seemingly minor actions taken early on, the junctures where lock-in processes begin to take hold can be seen as points at which artistic projects oriented toward environmental justice can productively intervene. Art can help pinpoint the moments when the corridors of the futures start to narrow, loosening self-reinforcing processes that lock in certain trajectories and exclude others.

This is where the second, more propositional mode makes itself felt in many of the artistic examples of this book, in not only "defending" future generations by identifying areas of imaginative entrapment that limit the capacity to act, but also providing new answers to the question of how the futures of environmental change can be imagined. While it is not uncommon for artworks to be described as 'pathbreaking'—especially in the context of a renewed interest in futurisms and the world-making qualities of science fiction—relatively few artistic projects interrogate the speculative infrastructures upon which they rely in such a way that loosens their grip. For creative practice to fulfill its promise as a tool for escaping imaginative lock-in, it must do more than endlessly generate representations of the futures of climate change and biodiversity loss, and instead engage in methodological experimentation, directing attention toward developing new tools of *futuring* that are conscious of their own power. Against a backdrop of "everything change" in which planetary conditions undermine predictions' purchase on *the* future, an opportunity presents itself to transform futures work from a specialized corporate, state, or academic pursuit into a rich field of chronopolitical experimentation and patterned creativity, where critique and proposition conjoin to make futures differently.

ACKNOWLEDGMENTS

The act of writing acknowledgments often occurs at a strange temporal juncture. While lists of names might be compiled before a book is finished, it's only after the bulk of work is done that it's possible to know exactly who to thank. Acknowledging is therefore largely an act of looking back, and in this case, looking back over the many years of stop-start activity that characterized the writing of this book. But writing these pages in late 2025 also carries with it a degree of anticipation. Future readers are conjured into existence before they have materialized, along with imagined effects as the book eventually makes its way into the world.

There is another sense in which writing these acknowledgments is tinged with anticipation. Like my first monograph initially drafted almost ten years ago, the manuscript for *Futures in the Making* was completed as I prepared to welcome a new human into the world. Even before they are born, this human provided inspiration, excitement, and added pressure to get the book done, all for which I'm truly grateful.

I'm lucky to say that some of the early readers of this book were also some of my dearest friends. A lot of the first draft was produced at the time of the Covid-19 pandemic, a period that would have been a lot harder if it weren't for the companionship of Josh Bowsher, David Eckersley, and many others. With Josh, I swapped numerous drafts and supportive chats, and with David, I spent many happy afternoons meandering through ideas in ways that are only possible on a long walk with an old friend.

The ideas that fed into these exchanges and later this book were made possible by an early career fellowship funded by the Leverhulme Trust from 2018 to 2022. Having the time to think at the start of an academic career is all too rare, and I'm grateful to the Trust and to my two research mentors on the fellowship—Jonathan Harris and Anthony Downey—for their support and generous advice through the ups and downs of this period. I have been lucky to work alongside many other wonderful colleagues at BCU since the book was conceived. Particular thanks go to Sian

Vaughan, whose informal mentoring and tireless work supporting others have cultivated the fertile soil from which so many research projects have grown. Past and present colleagues in the School of Art, especially Jenny Wright, John Wigley, Brian Bishop, and Lisa Metherell, have provided friendship, flexibility, and the priceless insights of seasoned practitioners over many years. Supervision meetings with my brilliant doctoral students Jo Callaghan, Laura Cooper, Pierre D'Alancaisez, Simon Fleury, Sophie Hedderwick, Joey Holder, Fred Hubble, and Federica Mirra constituted one of the most pleasurable parts of my day job, providing regular infusions of ideas and energy. Many of this group also brought their bodies and minds to the Environmental Humanities Study Group I set up at BCU, and I'm grateful for all the warm and generative conversations that characterized our small gatherings. In recent years, this group has morphed into the Lifeworlds Research Cluster in Art and Design, also hosted at BCU. Thanks go to all the people who have passed through, and a special mention to Kirsten Forkert for helping to formulate our first events series and Becky Shaw for her enthusiastic co-leadership in plotting paths forward across shifting institutional terrain.

There are many other groups that have nourished and sustained me these past seven years. What started as a series of informal conversations on political ecology with Yolanda Ariadne Collins later developed into a full-blown research project funded by the University of St. Andrews involving Lydia Cole and Matthew Barlow, all of whom became friends as well as collaborators. While the project sat in a different part of my brain and hard drive to the present book, I'm grateful for the many discussions, dinners, and one memorably icy swim off the coast of Scotland. I'm lucky to also be part of an extended network of early career researchers who met at the now defunct Centre for Cultural Studies (CCS) at Goldsmiths College, where I pursued doctoral research under the first-rate guidance of Matthew Fuller and Simon O'Sullivan. Over a decade later, I still benefit directly from their wisdom, and indirectly through the post-CCS research network held together by Nicole Sansone, Maria-Rosario Montero, and many others. Another group closer to home provided a venue for slow-burning conversations around art and ecology in the early days of this project. Initiated by Andrew Goffey, Lynn Pettinger, and Alba Colomo, the Three Ecologies Research Group was a welcome home for a diverse group of environmental thinkers and practitioners in my home city.

Early iterations of the research that eventually fed into this book have been presented at numerous conferences and symposia over the years. It's not possible to name all of the organizers, interlocutors, and audiences that have witnessed half-baked versions of the chapters, but some that particularly stood out were Christian Alonso and the team at MACBA Barcelona, Jorinde Seijdel and Jort van der Laan at the Gerrit Rietveld Academie in Amsterdam, Renata Tyszczuk at the University of Sheffield, David Sergeant at the University of Plymouth, Timotheus Vermeulen at the University of Oslo, Andreas Niegl and Yulia Gromova at the University of Kassel, David Bell and Laura Pulsgrove at Loughborough University, and Fernando García Dory and Amelie Aranguren at Inland—CAR Centro de Acercamiento a lo Rural in Madrid, for their generous hospitality while I was thinking through some ideas for the final chapter. Some of these ideas were also road-tested in a public event I organized at Nottingham Contemporary in 2022 titled *After Growth: A Symposium on Post-Capitalist Imaginaries*. Thanks go to Sam Thorne for the vote of confidence that enabled me to put on an event of this scale and ambition, to all of the speakers who brought their thoughts on art and degrowth to the table, to Sofia Lemos for her friendship and consummate organizational skills in the early planning stages, and to Canan Batur for her tireless energy bringing the event over the finish line.

Writing is known to be a famously solitary pursuit, often carried out in hushed libraries and private offices. As pleasurable as this can sometimes be, I'm lucky to have had the opportunity to break up periods of seclusion with access to a comparatively sociable workspace at Primary—an artist-led space that I share with a talented group of individuals. A particular shout out to friends and fellow studio holders Nicola Guy, Tom Harris, Roger Suckling, and the much-missed Alison Lloyd; Hugh Nicholson, Nastassja Simensky, Rebecca Beinart, Will Harvey, and Jacob Kelly, whose friendship and ability to bake political commitments into the core of their creative work is an inspiration.

Over its long period of gestation, the book has passed through several pairs of editorial hands. Caroline Picard, Giovanni Aloi, and Michael M. J. Fischer deserve a special mention for their encouragement in the early stages of the project. Jon K. Shaw served as a very capable midwife for the first full manuscript before it was handed over to the MIT Press. There, three anonymous reviewers engaged so closely and constructively with the draft that it restored my faith in the peer-review system. I'm grateful for Valeria Padalino for demonstrating how skillful indexing can be, Grace Sparapani for the copy editing, Judy Feldmann for managing the production process, and of course MIT Press's commissioning editor Gabriela Bueno Gibbs, who expertly handled the book through the various stages of its development.

A heartfelt thanks go to all the artists and creative practitioners discussed in the book, especially those who granted permission for me to use images of their work, in some cases even offering to fact check sections of writing (the ultimate responsibility for any remaining errors, it should be stressed, remains mine alone).

Finally, I want to mention the loving and supportive group of humans I'm lucky to call my family: Maggie and Barrie, Erica and Tom, Toby and Amy, Harriet and Alistair, my daughter Ada, for providing a constant source of hope for the future, and my wise and wonderful partner Ellie Higgins for her unwavering encouragement and the happiness our life together brings me every day.

Sections of chapter 1 first appeared as "The Art of Disciplined Imagination: Prediction, Scenarios, and Other Speculative Infrastructures," in *Critical Inquiry* 47, no. 4 (Summer 2021): 719–748, and an early version of chapter 2 has been previously published as "On Prospecting: Visual Culture between Extraction and Speculation," in *Environmental Humanities* 16, no. 3 (November 2024): 603–623. I'm grateful to the editors and peer reviewers of these articles for sharpening them up and making their republication possible.

NOTES

INTRODUCTION

1. IPCC, *Climate Change 2007: Impacts, Adaptation and Vulnerability. Contribution of Working Group II to the Fourth Assessment Report of the Intergovernmental Panel on Climate Change*, ed. Martin Parry et al. (Cambridge University Press, 2007), 145.
2. IPCC, *Climate Change 2007*, 144. In subsequent IPCC reports, the work of clarifying terminology around scenarios, projections, and other modes of characterizing the future has been left to glossaries. The choice to discuss this now rather old diagram is because it is one of the few places the authors synthesize this information into one image.
3. Projections are taken in the narrow sense here, as, in the words of the IPCC report, "a model-derived estimate of future conditions related to one element of an integrated system (e.g., an emission, a climate, or an economic growth projection)." IPCC, *Climate Change 2007*, 145.
4. Naomi Oreskes, "Why Believe a Computer? Models, Measures, and Meaning in the Natural World," in *The Earth Around Us: Maintaining A Livable Planet*, ed. Jill Schneiderman (Routledge, 2000), 70–82; Paul N. Edwards, *A Vast Machine: Computer Models, Climate Data, and the Politics of Global Warming* (MIT Press, 2010); Paul N. Edwards, "History of Climate Modeling," *WIREs Climate Change* 2 (2011): 128–139; J. R. Fleming, *Historical Perspectives on Climate Change* (Oxford University Press, 1998).
5. This is a concept I first developed in the article "The Art of Disciplined Imagination: Prediction, Scenarios, and Other Speculative Infrastructures," *Critical Inquiry* 47, no. 4 (2021): 719–748. For a good companion publication that also combines the words "infrastructure" and "speculation," albeit one rooted more in a more curatorial context, see Bassam El Baroni, ed., *Between the Material and the Possible: Infrastructural Re-examination and Speculation in Art* (Sternberg, 2022).
6. S. Beck and Jeroen Oomen, "Imagining the Corridor of Climate Mitigation—What Is at Stake in IPCC's Politics of Anticipation?" *Environmental Science and Policy* 123 (2021): 169–178. On "orchestration" as it relates to a range of predictive practices, see Eglė Rindzevičiūtė, *The Will to Predict: Orchestrating the Future through Science* (Cornell University Press, 2023), 9–10.
7. Bertrand de Jouvenel, *The Art of Conjecture* (Basic Books, 1967).
8. For an anthropological exploration of the operations of climate change models and their embeddedness within varied cultures and contexts, see Kirsten

Hastrup and Martin Skrydstrup, eds., *The Social Life of Climate Change Models: Anticipating Nature* (Routledge, 2013).

9. Ben Anderson, "Security and the Future: Anticipating the Event of Terror," *Geoforum* 41, no. 2 (2010): 229. See also Ben Anderson, "Preemption, Precaution, Preparedness: Anticipatory Action and Future Geographies," *Progress in Human Geography* 34, no. 6 (2010): 777–798. As will become clear over the course of this introduction, "anticipation" is being used here in a narrower sense than in the field of "anticipation studies," where the word covers "all 'ways of knowing' the later-than-now . . . from those forms of anticipation that are observed, for instance, in a tree that loses its leaves in the Autumn to human planning that attempts to colonize the future and efforts to make sense of emergent novelty in the present by finding inspiration in systemically discontinuous imaginary futures." Riel Miller, Roberto Poli, and Pierre Rossel, "The Discipline of Anticipation: Foundations for Futures Literacy," in *Transforming the Future: Anticipation in the 21st Century*, ed. Riel Miller (Routledge, 2018), 52. While this expanded use has been productive for a range of researchers, it can sometimes rub against the grain of common understandings of anticipation, which carry overtones of an expectation or confidence that something will happen, as in the sentence "She anticipated their arrival." The preferred general term here is "speculation." While it could be argued that this falls prey to the opposite problem, often being used to describe more exploratory futures work, when the realm of speculative finance is considered, it is able to accommodate a large repertoire of tools, techniques, and concepts, including those that structure the four chapters of this book. As mentioned later in the introduction, it is also preferred to "anticipation" for its etymological links to processes of vision and visualization.
10. Barbara Adam, "Memory of Futures," *KronoScope* 4, no. 2 (2004): 297–315.
11. While fears of nuclear catastrophe dominated the 1960s, these fears were often connected to narratives of biosphere collapse, or at least produced in parallel. Rachel Carson's *Silent Spring* (Houghton Mifflin, 1962) was an early contribution to the second camp, and in a different genre, Harlan Jay Ellison's "Boy and His Dog" (*New Worlds* 189 [April 1969]: 4–16) merged the two.
12. Wendy L. Schultz, "A Brief History of Futures," *World Futures Review* 7, no. 4 (2015): 324–331.
13. Rindzevičiūtė, *Will to Predict*; Jenny Andersson, *The Future of the World: Futurology, Futurists, and the Struggle for the Post–Cold War Imagination* (Oxford University Press, 2018).
14. On the establishment of this directory, the global futures network, and other aspect of the professionalization of the field, see Andersson, *Future of the World*, 211. It should be noted that while "futurist" as a professional title is still common, using "futurism" to describe the field is generally avoided on account of its associations with the early twentieth-century art movement. See Wendell Bell, *Foundations of Futures Studies* (Transaction, 1997), 69.
15. Toni Ahlqvist and Martin Rhisiart, "Emerging Pathways for Critical Futures Research: Changing Contexts and Impacts of Social Theory," *Futures* 71 (2015): 91–104.

16. Hyeonju Son, "The History of Western Futures Studies: An Exploration of the Intellectual Traditions and Three-Phase Periodization," *Futures* 66 (2015): 130.

17. Wendell Bell and James A. Mau, eds., *The Sociology of the Future: Theory, Cases and Annotated Bibliography* (Russell Sage Foundation, 1971); Michael Young, ed., *Forecasting and the Social Sciences* (Heinemann, 1968). Bell and others claimed that there was a future-oriented tradition to sociology since its inception, citing authors such as August Comte, Karl Marx, and Marquis de Condorcet.

18. Wendell Bell, "The Sociology of the Future and the Future of Sociology," *Sociological Perspectives* 39 (1999): 39–57.

19. Toward the later stages of writing this book I became aware that *Futures in the Making* shares its title with a concept Adam developed in her work independently and with Chris Groves. Here the authors use it to denote futures that are already underway, but not yet "congealed into matter" in forms available to sense perception. For Adam and Groves, the fact that such futures in the making have not yet materialized as symptoms does not imply that they are unreal. Drawing on a rich philosophical tradition from Baruch Spinoza to Jakob von Uexküll, the authors invite us to peel back the surface of the phenomenal world (what von Uexküll called the *Merkwelt*) and find ways of investigating the latent ongoing processes that effect reality (or *Wirkwelt*). Such processes cannot be experienced as material facts, but are nevertheless "set in motion by socio-political, legal, scientific, technological, economic, and everyday enacting practices." In this way Adam and Grove's concept of futures in the making has some overlaps with the idea of "infrastructural lock-in" I elaborate at various points in this book, most extensively in the conclusion. Barbara Adam, "Futures in the Making: Sociological Practice and Challenge," in *Handbook of Public Sociology*, ed. Vincent Jeffries (Rowman & Littlefield, 2009), 429–474 (429); Barbara Adam and Chris Groves, *Future Matters: Action, Knowledge, Ethics* (Brill, 2007).

20. Mads Borup et al., "The Sociology of Expectations in Science and Technology," *Technology Analysis and Strategic Management* 18, no. 3–4 (2006): 285–298; Harro van Lente, *Promising Technology: The Dynamics of Expectations in Technological Developments* (University of Twente, 1993); and Rebecca Coleman and Richard Tutton, "Futures in Question: Theories, Methods, Practices," special issue, *Sociological Review* 65, no. 3 (2017).

21. Roberto Poli, "The Subtle Interplay Between Futures Studies and Anticipation Theory," introduction to *Handbook of Futures Studies*, ed. Roberto Poli (Edward Elgar, 2024), 1. See also Roberto Poli, *Introduction to Anticipation Studies* (Springer 2017); and Robert Rosen, *Life Itself* (Columbia University Press, 1991). See note 9 above for elaboration on terminology.

22. Roberto Poli, "The Many Aspects of Anticipation," *Foresight* 12, no. 3 (2010): 8.

23. Riel Miller, "Futures Literacy: Transforming the Future," introduction to *Transforming the Future: Anticipation in the 21st Century*, ed. Riel Miller (Routledge, 2018), 5.

24. Ziauddin Sardar, "The Namesake: Futures; Futures Studies; Futurology; Futuristic; Foresight–What's in a Name?" *Futures* 42, no. 3 (2010): 180. See also John

Irvine and Ben R. Martin, *Foresight in Science: Picking the Winners* (Frances Pinter, 1984).

25. Son, "History of Western Futures Studies," 129. An important exception to this is the idea of "social foresight," pioneered by Richard Slaughter, which embodies an explicitly critical approach. Richard Slaughter, *Futures Beyond Dystopia: Creating Social Foresight* (Routledge, 2004).

26. Ahlqvist and Rhisiart, "Emerging Pathways for Critical Futures Research," 92; and Jan Erik Karlsen et al., "Sociological Contributions to Futures' Theory Building," *Foresight* 12, no. 3 (2010): 59–72.

27. During this period there are numerous figures associated with concepts aligned or adjacent to critical futures studies, for example, "critical futurology" (Bart van Steenbergen, "Critical and Establishment Futurology," in *Challenges from the Future: Proceedings of the International Future Research Conference*, ed. Japan Society of Futurology [Kodansha, 1970], 1:93–101); "critical rationalism" (Klaus Tuchel, "Social Philosophy in the Future: Aims and Problems," in *Challenges from the Future: Proceedings of the International Future Research Conference*, ed. Japan Society of Futurology [Kodansha Ltd, 1970], 3:17–23); and "public futures" (Ian Miles, *The Poverty of Prediction* [Saxon House, 1975]).

28. Alvin Toffler, *Future Shock* (Pan, 1971), 419 and 432. While the Tofflers's first three books bear Alvin's name alone, it is widely known that his wife Heidi made significant contributions. See Keith Schneider, "Heidi Toffler, Unsung Force Behind Husband's Futurist Books, Dies at 89," *New York Times*, February 13, 2019. For more recent work on the democratization of futures work and community-oriented projects aligned to this goal, see Denis Loveridge and Ozcan Saritas, "Reducing the Democratic Deficit in Institutional Foresight Programmes: A Case for Critical Systems Thinking in Nanotechnology," *Technological Forecasting and Social Change* 76, no. 9 (2009): 1208–1221; and Mihaela Ghişa et al., "Designing a Foresight Exercise for the Future of Rural Communities in Romania," *Futures* 43, no. 9 (2011): 996–1008.

29. For a critique of "ivory tower" futurism in the 1950s and '60s, see Johan Galtung, "On the Future of the International System," *Journal of Peace Research* 4, no. 4 (1967): 305–333.

30. As an example of tech firms providing market insights to customers, see the "Pinterest Predicts" trendspotting report: https://business.pinterest.com/en-gb/pinterest-predicts, accessed July 1, 2024. As an example of more general futures consultancy, Prognos AG is one of the oldest companies in the business. Founded in 1959, they now face competition from thousands of other futures and foresight companies. On governmental futures research, see the UK Government's horizon scanning and foresight projects: https://www.gov.uk/government/collections/foresight-projects, accessed July 1, 2024.

31. Stuart Candy, "Gaming Futures Literacy: The Thing from the Future," in *Transforming the Future Anticipation in the 21st Century*, ed. Riel Miller (Routledge, 2018).

32. Slaughter, *Futures Beyond Dystopia*, 175; Pamela McCorduck and Nancy Ramsey, *The Futures of Women: Scenarios for the 21st Century* (Addison-Wesley, 1996); Ulrika Gunnarsson-Östling, "Gender in Futures: A Study of Gender and Feminist Papers published in *Futures*, 1969–2009," *Futures* 43, no. 9 (2011): 1029–1039; and Ivana Milojević et al., eds., "Feminist Futures," special issue, *Futures* 40, no. 4 (2008).

33. Ahlqvist and Rhisiart, "Emerging Pathways for Critical Futures Research," 103.

34. Sardar, "Namesake," 183.

35. There has been an enormous amount written on the topic of speculative design in the past ten to fifteen years, stemming from the publication of books such as Anthony Dunne and Fiona Raby, *Speculative Everything: Design, Fiction, and Social Dreaming* (MIT Press, 2013); and Stuart Candy and Cher Potter, eds., *Design and Futures* (Tamkang University Press, 2019). At present, much less has been written on the future-making practices of visual art.

36. Matthias Heymann et al., introduction to *Cultures of Prediction in Atmospheric and Climate Science: Epistemic and Cultural Shifts in Computer-Based Modelling and Simulation*, ed. Matthias Heymann et al. (Routledge, 2017), 5.

37. Peter Galison, "The Future of Scenarios: State Science Fiction," in *The Subject of Rosi Braidotti: Politics and Concepts*, ed. Bolette Blaagaard and Iris van der Tuin (Bloomsbury Academic, 2014).

38. Slaughter, *Futures Beyond Dystopia*, 106; and Sohail Inayatullah, "Futures Studies: Theories and Methods," in *There's a Future: Visions for a Better World*, ed. Nayef Al-Fodhan et al. (TF, 2013), 44.

39. On this point, see also Richard Tutton, "Wicked Futures: Meaning, Matter and the Sociology of the Future," *Sociological Review* 65, no. 3 (2017): 487; Michael Godhe and Luke Goode, "Critical Future Studies," *Culture Unbound* 10, no. 2 (2018): 151–162; and Michael Godhe and Luke Goode, "Beyond Capitalist Realism—Why We Need Critical Future Studies," *Culture Unbound* 9, no. 1 (2017): 108–129.

40. Tutton, "Wicked Futures," 484.

41. Another artist investigating ocean infrastructure is Charles Lim Yi Yong, whose extensive body of work *SEA STATE* looks at Singapore's entanglement in diverse maritime histories and networks, including undersea internet cables. For a discussion of Lim's work, see Michael M. J. Fischer, *Probing Arts and Emergent Forms of Life* (Duke University Press, 2023).

42. Ruth Wilson Gilmore, "Abolition Geography and the Problem of Innocence," in *Futures of Black Radicalism*, ed. Gaye Theresa Johnson and Alex Lubin (Verso, 2016), 237. Gilmore's term riffs on "structures of feeling," a concept coined and developed by Raymond Williams in, among other places *The Long Revolution* (Chatto and Windus, 1961). As the following chapter shows, my own account repurposes the concept to consider the affective qualities of infrastructure.

43. *Oxford English Dictionary*, "speculation," https://www.oed.com/dictionary/speculation_n, accessed July 1, 2024.

44. See note 9 above for additional reasons "speculation" is preferred over "anticipation."

45. Macarena Gómez-Barris, *The Extractive Zone: Social Ecologies and Decolonial Perspectives* (Duke University Press, 2017), 133.

46. An early formulation of the notion of "operational images" appears in *Eye/Machine I*, dir. Harun Farocki (2000), video, color, 23 min. (loop). It has been subsequently developed in Volker Pantenburg, "Working Images: Harun Farocki and the Operational Image," in *Image Operations: Visual Media and Political Conflict*, ed. Jens Eder and Charlotte Klonk (Manchester University Press: 2016); and Jussi Parrika, *Operational Images: From the Visual to the Invisual* (University Minnesota Press, 2023).

47. T. J. Demos, *Beyond the World's End: Arts of Living at the Crossing* (Duke University Press, 2020), 5.

48. Demos, *Beyond the World's End*, 18.

49. Demos, *Beyond the World's End*, 18; Félix Guattari, *The Three Ecologies*, trans. Ian Pindar and Paul Sutton (Continuum, 2008).

50. Maarten A. Hajer and Peter Pelzer, "*2050—An Energetic Odyssey*: Understanding 'Techniques of Futuring' in the Transition Toward Renewable Energy," *Energy Research and Social Science* 44 (2018): 222–231.

51. Susan Leigh Star, "The Ethnography of Infrastructure," *American Behavioral Scientist* 43, no. 3 (1999): 381. See also Susan Leigh Star and Karen Ruhleder, "Steps Towards an Ecology of Infrastructure: Complex Problems in Design and Access for Large-Scale Collaborative Systems," in *CSCW '94: Proceedings of the 1994 ACM Conference on Computer Supported Cooperative Work* (Association for Computing Machinery, 1994).

52. Edwards, *Vast Machine*, 17. Many of Edwards's other writings explore knowledge infrastructure from different angles. See, for example, Paul N. Edwards, "Knowledge Infrastructures for the Anthropocene," *Anthropocene Review* 4, no. 1 (2016): 34–43.

53. As Mike Hulme has pointed out, in each of the IPCC report cycles completed to date, the report on "the physical science basis" has been published before more policy-orientated reports, thereby promoting a "science-led" model of public policy. Mike Hulme, "The IPCC's Mistaken 'Science-First' Approach to Climate Change," *Mike Hulme* (blog), August 9, 2021, https://mikehulme.org/the-ipccs-mistaken-science-first-approach-to-climate-change. The peer review process, also crucial as a means of ensuring trust, is taken very seriously by the IPCC, and authors are required to respond to each and every comment. For the working group contribution to the Sixth Assessment Report, published in August 2021, the reviewer comments totaled 74,000 in number.

54. A section of these might Michael Rubenstein et al., eds., "Infrastructuralism," special issue, *Modern Fiction Studies* 61, no. 4 (Winter 2015); Mary Bridges, "The Infrastructural Turn in Historical Scholarship," *Modern American History* 6, no. 1 (2023):103–120; Marina Vishmidt, "Beneath the Atelier, the Desert: Critique,

Institutional and Infrastructural," in *Marion von Osten: Once We Were Artists (A BAK Critical Reader in Artists' Practice)*, ed. Tom Holert and Maria Hlavajova (MIT Press and BAK, 2017); El Baroni, *Between the Material and the Possible*; and Lauren Berlant, "The Commons: Infrastructures for Troubling Times," *Environment and Planning D: Society and Space* 34, no. 3 (2016): 393–419.

55. Anne Spice, "Fighting Invasive Infrastructures: Indigenous Relations against Pipelines," *Environment and Society* 9 (2018): 40–56.

56. Elsewhere I have discussed this in terms of "disciplined" versus "undisciplined" imagination, repurposing a term from the management theorist Paul Schoemaker. See Theo Reeves-Evison, "The Art of Disciplined Imagination: Prediction, Scenarios, and Other Speculative Infrastructures," *Critical Inquiry* 47, no. 4 (2021): 719–748; and Paul J. H. Schoemaker, "Disciplined Imagination: From Scenarios to Strategic Options," *International Studies of Management & Organization* 27 (Summer 1997): 43–70.

57. Thomas J. Chermack and Susan A. Lynham, "Definitions and Outcome Variables of Scenario Planning," *Human Resource Development Review* 1, no. 3 (2002): 366–383; and Renata Tyszczuk, "A Brief History of Scenarios," in *Culture and Climate Change: Scenarios*, ed. Renata Tyszczuk et al. (Cambridge, 2014).

58. Jamie L. Pietruska, *Looking Forward: Prediction and Uncertainty in Modern America* (University of Chicago Press, 2017).

59. A small selection of this voluminous literature is as follows: Edwards, *Vast Machine* (2010); Paul N. Edwards "History of Climate Modeling," *Wiley Interdisciplinary Reviews: Climate Change* 2, no. 1 (2011): 128–139; Oreskes, "Why Believe a Computer?"; Hastrup and Skrydstrup, *Social Life of Climate Change Models*; Naomi Oreskes, "Changing the Mission: From the Cold War to Climate Change," in *Science and Technology in the Global Cold War*, ed. Naomi Oreskes and John Krige (MIT Press, 2014); Keynyn Brysse et al., "Climate Change Prediction: Erring on the Side of Least Drama?" *Global Environmental Change* 23, no. 1 (201): 327–337; Martin Mahony et al., "Cultures of Prediction in Climate Science" in *Climate and Culture: Multidisciplinary Perspectives on a Warming World*, ed. Giuseppe Feola et al. (Cambridge University Press, 2019); Matthias Heyman et al., eds., *Cultures of Prediction in Atmospheric and Climate Science: Epistemic and Cultural Shifts in Computer-Based Modelling and Simulation* (Routledge, 2017); Fleming, *Historical Perspectives on Climate Change*; and Spencer Weart, *The Discovery of Global Warming* (Harvard University Press, 2004).

60. It should be noted that the word "futurology" is not interchangeable with "futures studies." Several authors have sought to drive a wedge between the two: Jenny Andersson, for example, characterizes futurology as a mechanistic and control-oriented approach to the future rooted in Cold War science, and futures studies as a counter-project "that emphasized the imagination and the need to conjure peaceful and loving images of the future." See Andersson, *Future of the World*, 28. In reality the distinction is a lot more complex, exemplified by the fact that the term "futurology" was first coined by the Ossip Flechtheim in the 1940s. As a social democrat, humanist, and member of the German Society for Peace

and Conflict Studies, Flechtheim was very far from being an avatar of Cold War technoscience. See Ossip K. Flechtheim, "Teaching the Future," *Journal of Higher Education* 16, no. 9 (1945): 460–465. For a more detailed discussion of the nomenclature, see Sardar, "Namesake." For the most part this book avoids the word "futurology," unless in reference to historical and contemporary figures who use the word themselves.

61. Galison, "Future of Scenarios." It's worth noting that not-so-dry examples of science fiction have also played a part in military thinking, for example, the inclusion of Orson Scott Card's *Ender's Game* (Tor Books, 1985) on military reading lists, including those of the US Marine Corps.

62. Olaf Helmer, with contributions from Bernice Brown and Theodore Gordon, *Social Technology* (Basic, 1966). The term "predictive instrumentalities" features in a report published by the RAND Corporation. See Olaf Helmer and Nicholas Rescher, *On the Epistemology of the Inexact Sciences*, P-1513 (RAND, October 13, 1958). While it was not uncommon to see the predictions of the Cold War era clothed in the language of a systems rationality popular at the time, it is perhaps surprising to see military forecasting treated in a similar way more recently. See, for example, Ike Jeane, *Forecast and Solution: Grappling with the Nuclear, A Trilogy for Everyone* (Pocahontas, 1996): "The topic of war can be treated in some key regards as an objective science rather than as an exclusively subjective or normative one. . . . You plug in the values and you get the answer" (35).

63. For a recent reappraisal of the significance of the *Limits to Growth* report, see Ugo Bardi, "*The Limits to Growth*" *Revisited* (Springer, 2011). See also Donella Meadows et al., "*The Limits to Growth*": *The 30-Year Update* (Chelsea Green, 2004).

64. See, for example, the EU's large-scale "Destination Earth" project, which goes beyond the confines of most climate modeling and aims to create two "digital twins" of the earth. European Commission, "Destination Earth," https://digital-strategy.ec.europa.eu/en/policies/destination-earth, accessed July 1, 2024.

65. Brian Larkin, "The Politics and Poetics of Infrastructure," *Annual Review of Anthropology* 42 (August 2013): 329.

66. Stephen Graham and Nigel Thrift, "Out of Order: Understanding Repair and Maintenance," *Theory, Culture & Society* 24, no. 3 (2007): 8.

67. Michael Mann, "The Autonomous Power of the State: Its Origins, Mechanisms, and Results," *European Journal of Sociology / Archives Européennes de Sociologie / EuropäischesArchiv für Soziologie* 25, no. 2 (1984): 188.

68. Mann, "Autonomous Power of the State," 190. This dichotomy overlaps with Michel Foucault's widely discussed distinction between disciplinary power and governmentality.

69. Keller Easterling, *Extrastatecraft: The Power of Infrastructure Space* (Verso, 2014), 15.

70. Easterling, *Extrastatecraft*, 214.

71. The concept of "lock-in" first gained traction as a means to understand the particular challenges of moving away from fossil fuel infrastructure and the points

of resistance faced by those advocating for a just transition to alternative energy sources. For Gregory C. Unruh, "carbon lock-in" comprises several interdependent processes spread across technological systems and governing institutions that perpetuate the use of fossil fuel infrastructure even after alternatives have been shown to reduce both carbon emissions and costs. See Gregory C. Unruh, "Understanding Carbon Lock-In," *Energy Policy* 28 (2000): 817–830; and Gregory C. Unruh, "Escaping Carbon Lock-In," *Energy Policy* 30 (2002): 317–325. See also Antje Klitkou et al., "The Role of Lock-In Mechanisms in Transition Processes: The Case of Energy for Road Transport," *Environmental Innovation and Societal Transitions*, 16 (September 2015): 22–37; John Urry, *Societies Beyond Oil: Oil Dregs and Social Futures* (Zed, 2013); Philip J. Vergragt et al., "Carbon Capture and Storage, Bio-Energy with Carbon Capture and Storage, and the Escape from the Fossil-Fuel Lock-In," *Global Environmental Change* 21, no. 2 (May 2011): 282–292; and Matthias Kalkuhl et al., "Learning or Lock-In: Optimal Technology Policies to Support Mitigation," *Resource and Energy Economics* 34, no. 1(2012): 1–23.

As Helmrich et al. have shown, the concept of "lock-in" predates the term itself, often appearing under a variety of pseudonyms, such as "structural inertia," "social traps," and "entrenchment." See Alysha Helmrich et al., "Lock-In: Origination and Significance Within Infrastructure Systems," *Environmental Research: Infrastructure and Sustainability* 3, no. 3 (2023), https://doi.org/10.1088/2634-4505/acf7e6; Thomas Parker Hughes, *Networks of Power: Electrification in Western Society, 1880–1930* (Johns Hopkins University Press, 1983); Robert Costanza, "Social Traps and Environmental Policy," *Bioscience* 37 (1987): 407–412; and William Walker, "Entrapment in Large Technology Systems: Institutional Commitment and Power Relations," *Research Policy* 29, no. 7–8 (2000): 833–846.

72. Tutton, "Wicked Futures," 487.

73. See, for example, various contributions to Alex Wilkie et al., eds., *Speculative Research: The Lure of Possible Futures* (Routledge, 2017), as well as the work of Isabelle Stengers, Donna Haraway, and Didier Debaise, who have each reanimated discussions of speculation from the position of environmental theory and philosophy.

74. Donald Mckenzie, *An Engine, Not a Camera: How Financial Models Shape Markets* (MIT Press, 2006).

75. Borup et al., "Sociology of Expectations in Science and Technology," 292.

76. Poli, "Many Aspects of Anticipation," 12.

77. Hajer and Pelzer, "*2050—An Energetic Odyssey*"; and Tony Fry, *Defuturing: A New Design Philosophy* (Bloomsbury, 2020).

78. Andersson, *Future of the World*, 6.

79. See Tony Golding, *The City: Inside the Great Expectation Machine* (Pearson Education, 2001). While Golding coined the term to characterize the role of financial institutions in producing expectations, it could also be taken in a wider sense to cover a broader repertoire of tools, techniques, and methods that shape futures.

80. On scenarios as rehearsal spaces, see Renata Tyszczuk and Joe Smith, "Culture and Climate Change Scenarios: The Tole and Potential of the Arts and

Humanities in Responding to the '1.5 degrees target,'" *Current Opinion in Environmental Sustainability* 31 (April 2018): 56–64; and Beck and Oomen, "Imagining the Corridor of Climate Mitigation."

81. A 2021 IPCC report leaves little room for ambiguity in the framing of this fact, writing that "It is unequivocal that human influence has warmed the atmosphere, ocean and land," whereas in 2008 it was only "*likely* that there has been significant anthropogenic warming over the past 50 years averaged over each continent (except Antarctica)" (emphasis mine). See IPCC, "Summary for Policymakers," in *Climate Change 2021: The Physical Science Basis. Contribution of Working Group I to the Sixth Assessment Report of the Intergovernmental Panel on Climate Change*, ed. Valérie Masson-Delmotte et al. (IPCC and Cambridge University Press, 2021), 5; and IPCC, *Climate Change 2007: Synthesis Report. Contribution of Working Groups I, II and III to the Fourth Assessment Report of the Intergovernmental Panel on Climate Change*, ed. Core Writing Team et al. (IPCC, 2007), 5.

82. Mattias Hjerpe and Björn-Ola Linnér, "Utopian and Dystopian Thought in Climate Change Science and Policy," *Futures* 41, no. 4 (2009): 234–245.

83. The journal *Time and Society* has become a dynamic home for such conversations since it was established in the early 1990s.

84. Carol Greenhouse, *A Moment's Notice: Time Politics Across Cultures* (Cornell University Press, 1996).

85. Mark Rifkin, *Beyond Settler Time: Temporal Sovereignty and Indigenous Self-Determination* (Duke University Press, 2017).

86. Kyle Powys Whyte, "Time as Kinship," in *The Cambridge Companion to Environmental Humanities*, ed. Jeffrey Jerome Cohen and Stephanie Foote (Cambridge University Press, 2021).

87. The metaphor of climate change as a "ticking clock" or "ticking time bomb" is ubiquitous in the lexicon of mainstream politicians, notably featuring in UK Prime Minister Boris Johnson's opening speech to the Glasgow COP26 summit in 2021: "The clock is ticking to the furious rhythm of hundreds of billions of pistons and turbines and furnaces and engines with which we are pumping carbon into the air faster and faster. . . . We have the technology to deactivate that ticking doomsday device." See Prime Minister's Office, "PM address at COP26 World Leaders Summit Opening Ceremony," November 1, 2021, https://www.gov.uk/government/speeches/pm-address-at-cop26-world-leaders-summit-opening-ceremony.

88. Kyle P. Whyte, "Indigenous Science (Fiction) for the Anthropocene: Ancestral Dystopias and Fantasies of Climate Change Crises," *Environment and Planning E: Nature and Space* 1, no. 1–2 (2018): 224–242.

89. Lawrence R. Samuel, *Future: A Recent History* (University of Texas Press, 2009).

90. Daniel Rosenberg and Susan Harding, eds., *Histories of the Future* (Duke University Press, 2005).

91. François Hartog, *Regimes of Historicity. Presentism and the Experiences of Time* (Columbia University Press, 2015), xvii.

92. As Hartog acknowledges, this fundamental idea owes a great deal to the work of German historian Reinhart Koselleck. See Reinhart Koselleck, *Futures Past: On the Semantics of Historical Time* (Columbia University Press, 2004).

93. Hartog, *Regimes of Historicity*, xviii and 17–18.

94. Hartog, *Regimes of Historicity*, xviii.

95. Hartog, *Regimes of Historicity*, 139. In the preface to the English translation of *Regimes of Historicity* Hartog is at pains to resist this charge of nostalgia.

96. Chris Lorenz, "Out of Time? Some Critical Reflections on François Hartog's Presentism," in *Rethinking Historical Time: New Approaches to Presentism*, ed. Marek Tamm and Laurent Olivier (Bloomsbury Academic, 2019).

97. See also Franco "Bifo" Berardi, *After the Future* (AK, 2011).

98. In this way, it would be productive to read the work of Hartog and others through the "standpoint epistemology" of feminist science studies, as exemplified in Donna Haraway, "Situated Knowledges: The Science Question in Feminism and the Privilege of Partial Perspective," *Feminist Studies* 14, no. 3 (1988): 581.

99. Andreas Malm, *Fossil Capital: The Rise of Steam Power and the Roots of Global Warming* (Verso, 2016), 8.

100. See https://www.moorsforthefuture.org.uk, accessed July 1, 2024. I am grateful to the late Alison Lloyd for pointing me in the direction of these boggy facts.

101. Numerous other examples of the messy mix-up of environmental time scales can be found in the work of environmental humanities scholar Michelle Bastian, whose compelling reflections on multispecies time include considerations of the burgeoning populations of jellyfish and their impact on species such as the leatherback turtle: "The fear is that jellyfish might become so dominant that a regime shift could replace fish with jellyfish. . . . Relinquishing its implacable forward movement, time (whose time?) threatens to stall and begin to run in reverse, looping the Anthropocene back around into the Cambrian." See Michelle Bastian, "Encountering Leatherbacks in Multispecies Knots of Time," in *Extinction Studies: Stories of Time, Death and Generations*, ed. Thom van Doreen et al. (Columbia University Press, 2017), 165. See also Anthony J. Richardson et al., "The Jellyfish Joyride: Causes, Consequences and Management Responses to a More Gelatinous Future," *Trends in Ecology and Evolution* 24, no. 6 (2009): 312–322.

102. Vijay Kolinjivadi, "Can the Planet Really Be Saved in Time? On the Temporalities of Socionature, the Clock and the Limits Debate," *Environment and Planning E: Nature and Space* 3, no. 3 (2019): 911.

103. Céline Granjou et al., "The Politics of Anticipation: On Knowing and Governing Environmental Futures," *Futures* 92 (2017): 5–11; and Céline Granjou and Juan Francisco Salazar, "Future," *Environmental Humanities* 8, no. 2 (2016): 240–244.

104. See Melinda Cooper, "Turbulent Worlds: Financial Markets and Environmental Crisis," *Theory, Culture and Society* 27, no. 2 (2010): 167–190. For an artistic exploration of some of these instruments, see Gary Zhexi Zhang, ed., *Catastrophe Time* (Strange Attractor, 2023).

105. Barbara Adam, *Timescapes of Modernity: The Environment and Invisible Hazards* (Routledge, 1998), 90.

106. S. A. Levin, "Patchiness in Marine and Terrestrial Systems: From Individuals to Populations," *Philosophical Transactions of the Royal Society B: Biological Sciences* 343 (1994): 99–103. For an anthropological take on the concept of patchiness, see Anna Lowenhaupt Tsing et al., "Patchy Anthropocene: Landscape Structure, Multispecies History, and the Retooling of Anthropology: An Introduction to Supplement 20," *Current Anthropology* 60, S20 (2019): S186–S197.

107. The approach here could be aligned to the pragmatics of what ocean media scholar Melody Jue, in a different context, has called a "pre-activist" approach to scholarship, which provides new stories and analytical techniques as resources for others, or, stretching further back, a Deleuzean-Guattarian tradition of theoretical bricolage in which existing ideas can be disassembled and reassembled like artistic materials in such a way as to bring forth the new. See Melody Jue, *Wild Blue Media: Thinking Through Seawater* (Duke University Press, 2020), 33.

108. Michael J. Albert, *Navigating the Polycrisis: Mapping the Futures of Capitalism and the Earth* (MIT Press, 2024).

109. As an indicator of academic output, Ramírez and Selin note that the scenarios field generated some 2,200 peer-reviewed articles per year in 2014. See Rafael Ramírez and Cynthia Selin, "Plausibility and Probability in Scenario Planning," *Foresight* 16, no. 1 (2014): 54.

110. Sardar, "Namesake," 180.

111. James Ogilvy, *Futures Studies and the Human Sciences: The Case for Normative Scenarios* (Routledge, 1996).

112. Adam and Groves, *Future Matters*, xv.

CHAPTER 1

1. Johan Galtung, "On Future Research and Its Role in the World," in *Challenges from the Future: Proceedings of the International Future Research Conference*, vol. 1, compiled by the Japan Society of Futurology (Kodansha, 1970), 104.

2. Juanita Schläpfer-Miller and Manuela Dahinden, eds., *Climate Garden 2085: Handbook for a Public Experiment* (Park, 2017), 5.

3. In many ways, these outdoor plots constituted a separate experiment, carried out by the soil scientist Viviana Loaiza, who tilled the soil to varying depths representing different approaches to soil health. See Schläpfer-Miller and Dahinden, *Climate Garden 2085*, 46.

4. As Renata Tyszczuk notes, there has been some terminological confusion around climate scenarios in IPCC reports over the years, having been variously termed "predictions," "projections," and "forecasts," to more recently being presented as "pathways." See Renata Tyszczuk, "Collective Scenarios: Speculative Improvisations for the Anthropocene," *Futures* 134 (2021): 4. See also Dennis

Bray and Hans von Storch, "'Prediction' or 'Projection'? The Nomenclature of Climate Science," *Science Communication* 30, no. 4 (June 2009): 534–543.

5. Joseph Masco, "Bad Weather: On Planetary Crisis," *Social Studies of Science* 40, no. 1 (2010): 17.

6. Eglė Rindzevičiūtė's fascinating intellectual history of scientific prediction in the Soviet Union demonstrates that, while the cybernetic and systems analytical underpinnings of predictive rationality were largely Western in origin, debates about scientific prediction in the USSR co-evolved through dialogue, exchange, and creative adaptations with epistemologies developed in the rest of the world from at least the eighteenth century onwards. See Eglė Rindzevičiūtė, *The Will to Predict: Orchestrating the Future Through Science* (Cornell University Press, 2023).

7. Hermann Kahn, "A Methodological Framework: The Alternative World Futures Approach," in *The Essential Herman Kahn: In Defence of Thinking*, ed. Paul Dragos Aligica and Kenneth R. Weinstein (Lexington, 2009), 195.

8. Herman Kahn, "Alternative World Futures," paper HI-342-B IV, Hudson Institute, April 1964, quoted in Daniel Bell, "Twelve Modes of Prediction: A Preliminary Sorting of Approaches in the Social Sciences," *Daedalus* 93, no. 3 (1964): 865.

9. Bell, "Twelve Modes of Prediction," 866. The distinction between scenarios and predictions is regularly affirmed in future studies scholarship in the 1960s and 1970s, no doubt because the scenario planning still represented an unfamiliar mode of speculation to some readers at the time.

10. Schläpfer-Miller and Dahinden, *Climate Garden 2085*, 11 and 9.

11. Schläpfer-Miller and Dahinden's approach to science communication could be aligned to that of Massimiano Bucchi and Brian Trench, who define it as "the social conversation around science," rather than simply transferring information. See Massimiano Bucchi and Brian Trench, "Rethinking Science Communication as the Social Conversation around Science," *JCOM* 20, no. 3 (2021): 1–11; see also Sarah Davies and Maja Horst, "Science Communication as Culture: A Framework for Analysis," in *Routledge Handbook of Public Communication of Science and Technology*, 3rd ed., ed. Massimiano Bucchi and Brian Trench (Routledge, 2021).

12. Schläpfer-Miller and Dahinden, *Climate Garden 2085*, 5.

13. Suraje Dessai et al., "On the Role of Climate Scenarios for Adaptation Planning," *Global Environmental Change* 15, no. 2 (2005): 87–97.

14. Andrew Curry, "A Beginning: A Critical History of Scenarios," in *Routledge Handbook of Social Futures*, ed. Carlos López Galviz and Emily Spiers (Routledge, 2022); Ron Bradfield et al., "The Origins and Evolution of Scenario Techniques in Long Range Business Planning," *Futures* 37 (October 2005): 795–812. For a discussion of the ubiquitous use of scenarios in business thought in the run up to the Y2K computer bug, which takes in the Boston Consulting Group among other organizations, see Michael M. J. Fischer, "If Derrida Is the Gomez-Peña of Philosophy, What Are the Genres of Social Science? Just Gaming: The Y2k

Computer Bug and Other Uncertainties, a Critical Simulation Game," in *Para-Sites: A Casebook Against Cynical Reason*, ed. George E. Marcus (Chicago University Press, 2000).

15. As Elke Seefried shows, in the 1950s and '60s, the "futures field" remained largely oriented around an East–West axis. It was only on the 1970s that a more global perspective was adopted, partly in response to new ideas around interconnectivity drawn from systems analysis, dependency theories, and environmental discourse. See Elke Seefried, "Globalized Science: The 1970s Futures Field," *Centaurus*, 59, nos. 1–2 (February/May 2017): 40–57. See also Jenny Andersson, *The Future of the World: Futurology, Futurists, and the Struggle for the Post Cold War Imagination* (Oxford University Press, 2018).

16. Ruth Wilson Gilmore, "Abolition Geography and the Problem of Innocence," in *Futures of Black Radicalism*, ed. Gaye Theresa Johnson and Alex Lubin (Verso, 2016), 237.

17. Kodwo Eshun, "Further Considerations on Afrofuturism," *New Centennial Review* 3, no. 2 (2003): 297; Grace L. Dillon, ed., *Walking the Clouds: An Anthology of Indigenous Science Fiction* (University of Arizona Press, 2012); José Esteban Muñoz, *Cruising Utopia: The Then and There of Queer Futurity* (NYU Press, 2009); Robert McRuer, "The Then and There of Crip Futurity," *GLQ: A Journal of Lesbian and Gay Studies* 20, no. 4 (2014): 532–534; Gary Zhexi Zhang, "Where Next? Imagining the Dawn of the 'Chinese Century,'" *Frieze*, April 2017, https://www.frieze.com/article/where-next-imagining-dawn-chinese-century; Jussi Parikka, "Middle East and Other Futurisms: Imaginary Temporalities in Contemporary Art and Visual Culture," *Culture, Theory and Critique* 59 (2018): 40–58; T. J. Demos, *Radical Futurisms: Ecologies of Collapse, Chronopolitics, and Justice-to-Come* (Sternberg, 2023).

18. Eshun, "Further Considerations on Afrofuturism."

19. Renata Tyszczuk, "A Brief History of Scenarios," in *Culture and Climate Change: Scenarios*, ed. Renata Tyszczuk et al. (Shed, 2019); Matthew J. Spaniol and Nicholas J. Rowland, "Defining Scenario," *Futures & Foresight Science* 1, no. 1 (March 2019), https://doi.org/10.1002/ff02.3.

20. Dina Ternullo, *Characters and Scenarios of Early Commedia Dell'Arte*, no. 172 of *The Compleat Anachronist* (2016); Wolfgang U. Dressler, "'Scenario' as a Concept for the Functional Explanation of Language Change," in *Language Change and Functional Explanations*, ed. Jadranka Gvozdanovic (De Gruyter Mouton, 1997), 111.

21. It was nevertheless relatively common in the 1920s and '30s for multiple versions of the same film to be made simultaneously for different international audiences. See, for example, Georg Wilhelm Pabst's triple-headed version of *Don Quixote* (1933), produced in French, German, and English from the same scenario. I'm grateful to Jon K. Shaw for introducing me to this example.

22. While the scenario continues to fulfill its function of providing a brief summary of the work in progress, in early cinema, it sometimes played an additional role

in marketing the finished product too. See Isabelle Raynaud, "Screenwriting," in *The Encyclopedia of Early Cinema*, ed. Richard Abel (Routledge, 2005).

23. What typically differentiates the scenario from the screenplay is that the latter is written explicitly for the screen, whereas a scenario is a written story that can be understood independently of the medium. For more on the history of the scenario and its relationship to the screenplay and continuity script, see Steven Maras, *Screenwriting: History, Theory and Practice* (Wallflower, 2009), 89–92.

24. Thomas J. Chermack, and Susan A. Lynham, "Definitions and Outcome Variables of Scenario Planning," *Human Resource Development Review* 1, no. 3 (2002): 366–383. For all the emphasis on Kahn as the originator of scenario planning, it should be noted that a similar approach was developed contemporaneously in France by Bertrand de Jouvenel. Bringing together mainly French and American social scientists into an international forum to discuss methods for generating knowledge about the future, de Jouvenel coined the term "futuribles," which he defined as "a fan of possible futures," grounded in the idea that conjecture on possible alternatives should replace predictive certainty about the future. See Bertrand de Jouvenel, *The Art of Conjecture* (Basic, 1967), 18. See also Hugues de Jouvenel, "Futuribles: Origins, Philosophy, and Practices—Anticipation for Action," *World Futures Review* 11, no. 1 (2019): 8–18. It is notable that de Jouvenel also gave a talk at RAND in 1965: see Bertrand de Jouvenel, "FUTURIBLES," RAND Corporation, January 1965, 1, https://www.rand.org/content/dam/rand/pubs/papers/2008/P3045.pdf.

25. A claim made in Gill Ringland, *Scenario Planning: Managing for the Future* (John Wiley, 2006), 13–14. This is quite plausible, since Rosten wrote several successful screenplays in the 1940s and '50s, and was involved with the Social Sciences division at RAND following the war. Complicating this narrative is Harvey DeWeerd's claim that he heard the term 'scenario' used in a military context as early as 1944. See Harvey A. DeWeerd, "Political-Military Scenarios," paper presented at a conference on Modern Strategic Analysis held by the Security Studies Project of UCLA, January 13, 1967, and reproduced by the RAND Corporation, 2, https://www.rand.org/pubs/papers/P3535.html.

26. For a thorough discussion of the nuclear subject in literary fiction, see Paul Brians, *Nuclear Holocausts: Atomic War in Fiction, 1895–1964* (Kent State University Press, 1987).

27. Wiener quoted in Sharon Ghamari-Tabrizi, *The Worlds of Herman Kahn: The Intuitive Science of Thermonuclear War* (Harvard University Press, 2005), 76.

28. Herman Kahn and Anthony J. Wiener, "The Use of Scenarios," January 1, 1967, Hudson Institute, https://www.hudson.org/research/2214-the-use-of-scenarios.

29. Herman Kahn, *Thinking About the Unthinkable* (Princeton University Press, 1962).

30. Herman Kahn, *On Thermonuclear War* (Princeton University Press, 1960; new ed., Transaction, 2007), 416.

31. Ghamari-Tabrizi, *Worlds of Herman Kahn*, 203.

32. This decision to restrict release of *The War Game* was investigated in detail in a radio documentary in 2005, "The War Game Files," produced by David Morley for *Archive on 4*, 58 min., https://www.bbc.co.uk/programmes/b05xcktx. See also John R. Cook, "Who Banned *The War Game*? A Fifty-Year Controversy Reassessed," *Journal of British Cinema and Television* 14, no. 1 (2017): 39–63.

33. Williams points out that many of Kahn's scenarios are narrated in the present tense. This, I would argue, is primarily a rhetorical strategy to heighten dramatic effect. At its core, the logical structure of the scenario remains future-oriented and conditional, regardless of the language in which it is conveyed to a reader. R. John Williams, "World Futures," *Critical Inquiry* 42, no. 3 (Spring 2016): 522.

34. Peter Watkins, dir., *The War Game* (BBC, 1965), 5:03.

35. Watkins, *War Game*, 21:23. Another reference to Kahn is the inter-title at 40:35, "would the survivors envy the dead." "Will the Survivors Envy the Dead" is a chapter title from *On Thermonuclear War*. It is also widely thought that Kahn served as the model for Dr. Strangelove in Kubrick's film of the same name.

36. The installation was recreated for the 2002 exhibition *Iconoclash: Beyond the Image Wars in Science, Religion and Art*, curated by Peter Galison et al., Zentrum für Kunst und Medientechnologie, Karlsruhe, May 4–September 1, 2002.

37. Isozaki interviewed by Hans Ulrich Obrist, in Obrist, "Triennale di Milano 68. A Case Study and Beyond; Arata Isozaki's Electronic Labyrinths. A «MA» of Images," in *Iconoclash: Beyond the Image Wars in Science, Religion and Art*, exh. cat., ed. Bruno Latour and Peter Weibel (ZKW and MIT Press, 2002), 172.

38. Noboru Kawazoe, ed., *Metabolism 1960: The Proposals for a New Urbanism* (Bijutsu Shuppansha, 1960).

39. Arata Isozaki, "Ruins," in *Arata Isozaki*, ed. Ken Oshima, trans. John D. Lamb (Phaidon, 2009), 29.

40. It's worth noting that later Isozaki came to regret his participation in the Expo, likening the mobilization of nationalist sentiment it helped bring about to participating in a war. Isozaki Arata, *Kūkan-e* [Toward Space] (Bijutsu Shuppansha, 1971), 511, quoted in Hyunjung Cho, "Expo '70: The Model City of an Information Society," *Review of Japanese Culture and Society* 23 (2011): 58.

41. The first International Future Research Conference, "Mankind 2000," held in Oslo in 1967 was a much more diverse affair, as Jenny Andersson notes, involving both "civilian defence intellectuals" rooted in Cold War think tanks and many from the new left and peace movements, as well as speakers from the USSR. See Robert Jungk and Johan Galtung, eds., *Mankind 2000* (PRIO, 1968).

42. To add further connections, science fiction novelist Sakyo Komatsu and architectural critic Noboru Kawazoe were both Metabolist collaborators, active in shaping the Expo and involved in establishing futures studies in Japan. For more on the creative, ideological, and organizational overlaps, see William O. Gardner, "Narratives of Collapse and Generation: Komatsu Sakyō's Disaster Novels and the Metabolist Movement," *Japan Forum* 26, no. 3 (2014): 306–324;

William O. Gardner, "The 1970 Osaka Expo and/as Science Fiction," *Review of Japanese Culture And Society* 23 (2011): 26–43; Cho, "Expo '70"; Yujiro Hayashi, "Futures Research in Japan," in *Handbook of Futures Research*, ed. Jib Fowles (Greenwood, 1978), 41.

43. Gardner, "Narratives of Collapse and Generation."

44. Johan Galtung, "On Future Research and Its Role in the World," 105.

45. For a more contemporary example of the influence of Hollywood on scenario practices, see Grégoire Mallard and Andrew Lakoff, "How Claims to Know the Future Are Used to Understand the Present," in *Social Knowledge in the Making*, ed. Charles Camic et al. (Chicago University Press, 2012). This chapter details a large-scale scenario exercise called "Dark Winter," held in June 2001, that simulated a smallpox attack on the US. It is striking how much of the imagery used in this scenario exercise, including simulated news footage of the attack, is articulated in a visual language almost indistinguishable from those in disaster films.

46. On Shchedrovitsky and alternatives to Soviet centralized planning see Eglė Rindzevičiūtė, "The Future as an Intellectual Technology in the Soviet Union: From Centralised Planning to Reflexive Management," *Cahiers Du Monde Russe* 56, no. 1 (2015): 111–134. Shchedrovitsky formulated the idea of "forward reflexiveness," which departed from scientific forecasting in a similar way to scenario planning's break with predictive rationality. The tradition is nevertheless distinctive, and through simulation games or what he called "activity-thinking," Shchedrovitsky staged training exercises that could be performative, durational, and experiential in their manifestation, with one simulation game in 1979 lasting twenty-five days.

47. Bell, "Twelve Modes of Prediction," 869.

48. DeWeerd, "Political-Military Scenarios," 4.

49. At base, the Delphi method is a way of structuring questions and answers into a series of "rounds" where a number of experts are encouraged to offer their predictions independently and then refine their judgments based on the opinions of others. I have written about the cultural dimensions of the Delphi method in "The Art of Disciplined Imagination: Prediction, Scenarios, and Other Speculative Infrastructures," *Critical Inquiry* 47, no. 4 (2021): 719–748. For an account of an early application of the Delphi method, see Abraham Kaplan et al., "The Prediction of Social and Technological Events," *Public Opinion Quarterly* 14 (1950): 93–110.

50. *Toward the Year 2000: Work in Progress*, 30th anniv. ed., ed. Daniel Bell and Stephen R. Graubard (MIT Press, 1997).

51. Neil Pickett, *A History of Hudson Institute, 1961–1992* (Hudson Institute, 1992).

52. Williams, "World Futures," 524. See also B. Bruce Briggs, *Supergenius: The Mega-Worlds of Herman Kahn* (Lulu, 2000), 304–305, for a full list.

53. Fischer, "If Derrida Is the Gomez-Peña of Philosophy"; Andrew Curry, "A Beginning: A Critical History of Scenarios," in *Routledge Handbook of Social Futures*, ed. Carlos López Galviz and Emily Spiers (Routledge, 2022).

54. Melinda Cooper, "Turbulent Worlds: Financial Markets and Environmental Crisis," *Theory, Culture and Society* 27, no. 2 (2010): 167–190. The meteorological component of Cooper's article deals with the growth of financial instruments such as weather derivatives.

55. Cooper, "Turbulent Worlds," 184.

56. Angela Wilkinson and Roland Kupers, "Living in the Futures," *Harvard Business Review* 91, no. 5 (2013): 1.

57. Thomas J. Chermack, *Foundations of Scenario Planning: The Story of Pierre Wack* (Routledge, 2017).

58. Kees van de Heidjen, *Scenarios: The Art of Strategic Conversation*, 2nd ed. (John Wiley, 2005), 5–6. The scenario that described the refusal of oil-producing nations to continue to increase the oil supply was initially one of six discussed at Shell, known as the "crisis scenario." See also Wilkinson and Kupers, "Living in the Futures."

59. Joe Willmore, "Scenario Planning: Creating Strategy for Uncertain Times," *Information Outlook* 5 (2001): 9.

60. Van de Heidjen, *Scenarios*, 6. Or, to use Pierre Wack's phrase, which provides the title for the previous section, scenario planning can be characterized as "the gentle art of reperceiving." Pierre Wack, "Scenarios: Shooting the Rapids," *Harvard Business Review*, November 1985, https://hbr.org/1985/11/scenarios-shooting-the-rapids.

61. Pierre Wack, "Scenarios: Uncharted Waters Ahead," *Harvard Business Review* 63, no. 5 (1985): 73–89.

62. Frithjof Kublik et al., *Shell Energy Scenarios: Germany* (Royal Dutch Shell, 2017), 3.

63. Clement Bezold and Trevor Hancock, "An Overview of the Health Futures Field for the WHO Health Futures Consultation," in Ann Taket, *Health Futures in Support of Health for All: Report of an International Consultation, Geneva, 19–23 July 1993* (World Health Organization, 1993).

64. Clem Sunter quoted in Chermack, *Foundations of Scenario Planning*, 146. A similar claim is made by Stewart Brand in *The Clock of the Long Now: Time and Responsibility* (Basic, 1999).

65. Demos, *Radical Futurisms*, 40. In calling for a "future of many futures," this argument deliberately references the Zapatista demand for "a world in which many worlds fit," as well as subsequent literature drawing on pluriverse thinking, such as Marisol de la Cadena and Mario Blaser, eds., *A World of Many Worlds* (Duke University Press, 2018).

66. Demos, *Radical Futurisms*, 128. The argument for the radicalism of temporal heterogeneity also unravels when it comes into contact with recent corporate strategies in temporal design, such as Facebook's "Flick," a unit of time designed for video editing across different frame rates, and a technique developed by Google for adding leap seconds. See Larissa Pschetz et al., "Revealing Social

Infrastructures of Time," in *Speaking for the Social: A Catalogue of Methods*, ed. Hannah Knox and Gemma John (Punctum, 2022).

67. Robert Jungk, preface to *Mankind 2000*, ed. Robert Jungk and Johan Galtung (PRIO, 1968), i.

68. Johan Galtung, "On the Future of the International System," in *Mankind 2000*, ed. Robert Jungk and Johan Galtung (PRIO, 1968); Galtung, "On Future Research and Its Role in the World," 107.

69. Fred Polak, *The Images of the Future*, trans. Elise Boulding (Elsevier, 1973); Kenneth Boulding, *The Image: Knowledge in Life and Society* (University of Michigan Press, 1956); Elise Boulding and Kenneth Boulding, *The Future: Images and Processes* (Sage, 1995).

70. Andersson, *The Future of the World*, 165.

71. Such utopian ambitions does not mean that the images futures researchers used were always positive, as can be seen from the consciousness-raising doomsday scenarios of the 1970s and '80s such as *Limits to Growth* and the nuclear winter scenarios produced by the "TTAPS" research group.

72. On this see Williams, "World Futures," which emphasizes the link between scenario planning and what he labels "Oriental Systems Theory," understood as a mixed bag of ideas drawn from Eastern religious traditions. Wack's lifelong interest in spiritual traditions, from Zen Buddhism to the mystical teachings of G. I. Gurdjieff, has been well documented. See Chermack, *Foundations of Scenario Planning*; Richard Barbrook and Andy Cameron, "The Californian Ideology," *Science as Culture* 6, no. 1 (1996): 44–72; Fred Turner, *From Counterculture to Cyberculture* (University of Chicago Press, 2006). The Global Business Network would be instrumental in this later part of the story. Located in San Francisco, its countercultural credentials were underpinned by figures such as Stewart Brand while it expanded its portfolio of clients to include multinationals and state departments.

73. Another application of scenario planning by environmental campaigners is highlighted by Stefan Aykut, who discusses their use in a campaign that emerged from the anti-nuclear movement in the 1980s. Aykut speaks of scenarios as "strategic devices" that allowed "for the representation of contrasting alternative energy futures, thereby re-politicizing the energy debate." See Stefan Cihan Aykut, "Energy Futures from the Social Market Economy to the *Energiewende*: The Politicization of West German Energy Debates, 1950–1990," in *The Struggle for the Long-Term in Transnational Science and Politics*, ed. Jenny Andersson and Eglė Rindzevičiūtė (Routledge, 2015), 64.

74. Toni Ahlqvist and Martin Rhisiart, "Emerging Pathways for Critical Futures Research: Changing Contexts and Impacts of Social Theory," *Futures* 71 (August 2015): 91–104.

75. van de Heidjen, *Scenarios*, 9.

76. At the time of writing, Shell was in the process of trying to sell its Niger Delta business to Renaissance Africa Energy, a deal that many civil society organizations

worry will enable the company to walk away from the region without paying for clean-up costs or compensation for those affected by the activities of the business. See Amnesty International, "Nigeria: Government Must Halt Shell's Sale of Its Niger Delta Business Unless Human Rights Are Fully Protected," April 15, 2024, https://www.amnesty.org/en/latest/news/2024/04/nigeria-government-must-halt-shells-sale-of-its-niger-delta-business-unless-human-rights-are-fully-protected.

77. In 2009, a legal case dealing with Shell's human rights abuses against the Ogoni people reached an out-of-court settlement with the Saro-Wiwa family for 15.5 million US dollars. See Jad Mouawad, "Shell to Pay $15.5 Million to Settle Nigerian Case," *New York Times*, June 8, 2009, https://www.nytimes.com/2009/06/09/business/global/09shell.html?ref=global.

78. Damian Carrington and Jelmer Mommers, "'Shell Knew': Oil Giant's 1991 Film Warned of Climate Change Danger," *Guardian*, February 28, 2017, https://www.theguardian.com/environment/2017/feb/28/shell-knew-oil-giants-1991-film-warned-climate-change-danger.

79. The discovery of the greenhouse effect was made significantly earlier, in 1856, by Eunice Foote in a paper presented at the American Association for the Advancement of Science.

80. According to Newton Harrison, the map that forms the basis of the piece was one used by pilots, with all written information removed. Newton Harrison, interview with the author, November 26, 2020. The particular map projection would seem to be the Lambert azimuthal equal-area projection, typically used in scientific fields such as geology.

81. In 1959, Plass predicted that CO^2 levels in 2000 would be 30 percent higher than those in 1900 with an accompanying 1 degree Celsius rise in average temperature. Almost fifty years later, the Fourth IPPC Assessment Report estimated that there had been a 37 percent rise in CO^2 levels over the same period, with a rise in temperature of around 0.7 degrees Celsius. See Gilbert N. Plass et al., "Carbon Dioxide and the Climate," *American Scientist* 98, no. 1 (2010): 58–62. Early publications from Plass on climate change include Gilbert N. Plass, "The Carbon Dioxide Theory of Climate Change," *Tellus* 8, no. 2 (1956): 140–154; and Gilbert N. Plass, "The Influence of the 15u Carbon-Dioxide Band on the Atmospheric Infra-Red Cooling Rate," *Quarterly Journal of the Royal Meteorological Society* 82, no. 353 (1956): 310–324.

82. In their early work, the Harrisons often embedded a systems-based vision of ecology that involved stripped down ecosystems installed in gallery settings. See Helen Mayer Harrison and Newton Harrison, *The Time of the Force Majeure: After 45 Years, Counterforce Is on the Horizon*, ed. Petra Kruse and Kai Reschke (Prestel, 2016).

83. Harrison and Harrison, *Time of the Force Majeure*, 50.

84. Hermann Flohn, "Climate and Energy: A Scenario to a 21st Century Problem," *Climatic Change* 1 (1977): 5–20.

85. Flohn, "Climate and Energy," 10.

86. Mike Hulme, "The First Climate Scenario: A Drama in Three Acts," in *Culture and Climate Change: Scenarios*, ed. Renata Tyszczuk et al. (Shed, 2019).

87. Flohn, "Climate and Energy," 10.

88. The etymology of aesthetics is typically traced to the ancient Greek word *aisthētikos*, meaning "of sense perception."

89. Flohn, "Climate and Energy," 12.

90. Collyns quoted in Wilkinson and Kupers, "Living in the Futures," 8; Andersson, *Future of the World*, 196.

91. Donella H. Meadows et al., *The Limits to Growth: A Report for the Club of Rome's Project on The Predicament of Mankind* (Universe, 1972). Insofar as *Limits to Growth* applied the World3 model to simulate changes to five distinct areas of human activity—population, food production, industrial production, pollution, and consumption of non-renewable natural resources—many have described it using a combination of modeling and scenario planning, despite there being no mention of the latter in the report itself. See, for example, Lauren Rickards et al., "Opening and Closing the Future: Climate Change, Adaptation, and Scenario Planning," *Environment and Planning C: Government and Policy* 32, no. 4 (2014): 587–602; and Gerald O. Barney, *The Global 2000 Report to the President of the U.S.* (Pergamon, 1980).

92. Gunnar Myrdal, *Against the Stream—Critical Essays on Economics* (Vintage, 1975), 204–205, quoted in Julian L. Simon and Herman Kahn, introduction to *The Resourceful Earth: A Response to Global 2000*, ed. Julian L. Simon and Herman Kahn (Blackwell, 1984), 34–35.

93. This criticism has also been made by geographers and sociologists. See Rickards et al., "Opening and Closing the Future," 598; and Mattias Hjerpe M and Björn-Ola Linnér, "Utopian and Dystopian Thought in Climate Change Science and Policy," *Futures* 41, no. 4 (May 2009): 234–245. Rickards et al. note that the IPCC has also been known to enlist recognized scenario experts to bolster the credibility of their scenarios, including some based within Shell (596).

94. Rickards et al., "Opening and Closing the Future," 597; Dessai et al., "On the Role of Climate Scenarios for Adaptation Planning."

95. Mike Hulme and Suraje Dessai, "Negotiating Future Climates for Public Policy: A Critical Assessment of the Development of Climate Scenarios for the UK," *Environmental Science and Policy* 11, no. 1 (2008): 64.

96. Newton Harrison, interview with the author, November 25, 2020.

97. Harrison and Harrison, *Time of the Force Majeure*, 344.

98. Harrison and Harrison, *Time of the Force Majeure*, 346.

99. Galtung, "On Future Research and Its Role in the World," 109.

100. Tyszczuk, "A Brief History of Scenarios," 22. See also Renata Tyszczuk and Joe Smith, "Culture and Climate Change Scenarios: The Role and Potential of the

Arts and Humanities in Responding to the '1.5 Degrees Target,'" *Current Opinion in Environmental Sustainability* 31 (April 2018): 56–64.

101. Kathryn Yusoff and Jennifer Gabrys, "Climate Change and the Imagination," *Wiley Interdisciplinary Reviews: Climate Change* 2, no. 4 (2011): 516–534.

102. The project emerged from the residency program *Culture and Climate Change: Scenarios*, mentioned in earlier footnotes. See https://www.futurescenarios.co.uk.

103. See, for example, Jennifer Wenzel, introduction to *Fueling Culture*, ed. Imre Szeman (Fordham University Press, 2017); and Jeff Diamanti, *Climate and Capital in the Age of Petroleum: Locating Terminal Landscapes* (Bloomsbury Academic, 2021).

104. Peter U. Clark et al., "Consequences of Twenty-First-Century Policy for Multi-Millennial Climate and Sea-Level Change," *Nature Climate Change* 6, no. 4 (2016): 360. As Adreas Malm points out, "2100, the year where most scenarios and projections abruptly end . . . has no real terminal status. The widespread usage of benchmark is an accident of computer technology, early models having been unable to carry scientists any further." See Andreas Malm, *The Progress of This Storm: Nature and Society in a Warming World* (Verso, 2018), 6.

105. Lena Dobrowolska and Teo Ormond-Skeaping, "Interview with Kozlowski Kris Moore," *Palm*, June 2019. https://palmstudios.co.uk/feature/lena-dobrowolska-teo-ormond-skeaping.

106. William Gibson, interview with Brooke Gladstone, *Talk of the Nation*, National Public Radio, November 30 1999, https://www.npr.org/2018/10/22/1067220/the-science-in-science-fiction. For more on the attribution of this quote, see "Quote Origin: The Future Has Arrived—It's Just Not Evenly Distributed Yet," Quote Investigator, https://quoteinvestigator.com/2012/01/24/future-has-arrived.

107. Paul J. H. Schoemaker, "Disciplined Imagination: From Scenarios to Strategic Options," *International Studies of Management & Organization* 27, no. 2 (Summer 1997): 43–70.

108. Reeves-Evison, "The Art of Disciplined Imagination"; Hyeonju Son, "The History of Western Futures Studies: An Exploration of the Intellectual Traditions and Three-Phase Periodization," *Futures* 66 (February 2015): 120–137.

109. Not to be confused with the Futures Industry Association, a trade association for brokers on the US futures exchanges. Eshun, "Further Considerations on Afrofuturism," 290.

110. Eshun, "Further Considerations on Afrofuturism," 292.

111. Eshun, "Further Considerations on Afrofuturism," 292 and 293.

112. Eshun, "Further Considerations on Afrofuturism," 297.

113. Ahlqvist and Rhisiart, "Emerging Pathways for Critical Futures Research." Here the borders between state and non-state actors become rather blurry, as neoliberal organizations such as the IMF and the World Bank become the central protagonists in binding nations to this developmental trajectory.

114. Bart van Steenbergen, "Critical and Establishment Futurology," in *Challenges from the Future: Proceedings of the International Future Research Conference* (Kodansha, 1970).
115. Andersson, *Future of the World*, 154.
116. Ibrahim El-Essawy, "Futures Studies in Egypt," *Arab Economic Journal* 19 (2000): 3–17.
117. Ashis Nandy, "Bearing Witness to the Future," *Futures* 28, no. 6–7 (1996): 636–639; Wendy L. Schultz, "A Brief History of Futures," *World Futures Review* 7, no. 4 (2015): 331.
118. Black Quantum Futurism, *Space-Time Collapse II: Community Futurisms* (Afrofuturist Affair / House of Future Sciences Books, 2020), 47.
119. Sofia Samatar, "Toward a Planetary History of Afrofuturism," *Research in African Literatures* 48, no. 4 (Winter 2017): 187.
120. Grace L. Dillon, "Imagining Indigenous Futurisms," introduction to *Walking the Clouds: An Anthology of Indigenous Science Fiction*, ed. Grace L. Dillon (University of Arizona Press, 2012), 3.
121. Anaïs Duplan, "Moor Mother Explains Black Quantum Futurism," *Vice*, March 14, 2017, https://www.vice.com/en/article/ae8yye/moor-mother-explains-black-quantum-futurism.
122. Clara Herrmann, "In Conversation with Rasheedah Phillips: Black Futurism & Technologies of Joy," *Schloss Post* 0 (August 2018), https://schloss-post.com/black-futurism-technologies-joy.
123. Perhaps the most widely cited of these in recent years is Anna Tsing's concept of the plantationocene. See also Jaskiran Dhillon, introduction to "Resurgence, Decolonialization, and Movements for Environmental Justice," ed. Jaskiran Dhillon, special issue, *Environment and Society* 9, no. 1 (2018): 1–5; and Malcom Ferdinand, *Decolonial Ecology: Thinking from the Caribbean World* (Polity, 2021).
124. Herrmann, "In Conversation with Rasheedah Phillips."

CHAPTER 2

1. Bruno Latour, "An Attempt at a 'Compositionist Manifesto,'" *New Literary History* 41, no. 3 (2010): 486. Copyright © 2010 New Literary History, The University of Virginia. Reprinted with permission of Johns Hopkins University Press.
2. Toby Jurovics, "Framing the West: The Survey Photographs of Timothy H. O'Sullivan," in Toby Jurovics et al., *Framing the West: The Survey Photographs of Timothy H. O'Sullivan* (Yale University Press, 2010), 15. This number does not include support personnel or the group's military escort.
3. Clarence King et al., *Report of the Geological Exploration of the Fortieth Parallel*, vol. 3, *Mining Industry* (Professional Papers of the Engineer Department, US Army, 1870), vii.

4. From 1871 to 1874, O'Sullivan also took part in another survey led by George M. Wheeler, whose aims included gathering information on "the numbers, habits, and disposition of the Indians . . . , sites as may be of use for future military operations or occupation, and the facilities offered for making rail or common roads." Letter from Andrew A. Humphreys to George M. Wheeler, March 23, 1871, in A. A. Humphreys, *Preliminary Report Concerning Explorations and Surveys Principally in Nevada and Arizona* (US Government Printing Office, 1872), 11–12.

5. The instrumentalization of geological knowledge is evident in the field of economic geology, which took shape towards the end of the nineteenth century. Today, it lays claim to being a "predictive science" that has the capacity to anticipate the location of undiscovered mineral deposits through sophisticated technologies including machine learning. Michael Jébrak, "Economic Geology: Then and Now," *Geoscience Canada* 33, no. 2 (2006): 81–93.

6. While the word "bioprospecting" is relatively new, its practices stretch at least as far back to the advent of colonial botany in the early modern period. See Londa Schiebinger and Claudia Swan, eds., *Colonial Botany: Science, Commerce, and Politics in the Early Modern World* (University of Pennsylvania Press, 2005).

7. The word "prospect" can of course carry many other meanings, including that of a candidate: someone or something that may be suitable for a future role or purpose.

8. Macarena Gómez-Barris, *The Extractive Zone: Social Ecologies and Decolonial Perspectives* (Duke University Press, 2017), 133.

9. Donna Haraway, "Situated Knowledges: The Science Question in Feminism and the Privilege of Partial Perspective," *Feminist Studies* 14, no. 3 (1988): 575–599.

10. A selection of this literature includes Paul Virilio, *Speed and Politics: An Essay on Dromology* (Semiotext(e): 1977); Beau Grosscup, *Strategic Terror: The Politics and Ethics of Aerial Bombardment* (Zed, 2006); Derek Gregory, "Lines of Descent," *Open Democracy*, November 8, 2011, http://www.opendemocracy.net/derek-gregory/lines-of-descent; Eyal Weizman, "The Politics of Verticality: The West Bank as an Architectural Construction," in *Territories: Islands, Camps, and Other States of Utopia*, ed. Anselm Franke (Walther König, 2003); Eyal Weizman, *Hollow Land: Israel's Architecture of Occupation* (Verso, 2007); Stuart Elden, "Secure the Volume: Vertical Geopolitics and the Depth of Power," *Political Geography* 34 (2013): 35–51; Hanna Rose Shell, *Hide and Seek: Camouflage, Photography, and the Media of Reconnaissance* (Zone, 2012); Peter Adey et al., *From Above: War, Violence, and Verticality* (Hurst, 2013); and Caren Kaplan, *Aerial Aftermaths: Wartime from Above* (Duke University Press, 2018).

11. Denis Cosgrove, *Apollo's Eye: A Cartographic Genealogy of the Earth in the Western Imagination* (Johns Hopkins University Press, 2001).

12. James C. Scott, *Seeing Like a State: How Certain Schemes to Improve the Human Condition Have Failed* (Yale University Press, 1998).

13. For a small selection of material from the *Whole Earth Catalog*, see Caroline Maniaque-Benton with Meredith Gaglio, eds., *Whole Earth Field Guide* (MIT Press, 2016).

14. In order of appearance: Haraway, "Situated Knowledges"; Mary Louise Pratt, *Imperial Eyes: Travel Writing and Transculturation* (Routledge, 2008); Gómez-Barris, *Extractive Zone*.

15. T. J. Demos, *Against the Anthropocene: Visual Culture and the Environment Today* (Sternberg, 2017), 61–64.

16. Richard Misrach and Kate Orff, *Petrochemical America* (Aperture, 2012).

17. Demos, *Against the Anthropocene*, 71.

18. Gómez-Barris, *Extractive Zone*, 6.

19. Heather Houser, *Infowhelm: Environmental Art and Literature in an Age of Data* (Columbia University Press, 2020), 178.

20. Houser, *Infowhelm*, 198.

21. In 2020, lithium was added to the European Union's list of critical raw materials for the first time. For a compelling example of how artists might engage with the subject of critical raw materials, see the project "Euro—Vision" by the art duo FRAUD (Audrey Samson and Francisco Gallardo). Project website available at https://euro-vision.net.

22. *Charging Myths*, curated by Ils Huygens and On-Trade-Off, Z33 House for Contemporary Art, Design & Architecture, Hasselt, Belgium, March 6–August 21, 2022.

23. The original spelling of the name in the title of the series has been modified from Barsin to Barzin for consistency.

24. Georges Senga, interview with the author, June 8, 2022. See also Miles Larmer, *Living for the City: Social Change and Knowledge Production in the Central African Copperbelt* (Cambridge University Press, 2021), 40; Karel Van Nieuwenhuyse, *Tussen buit en baat: Congo in het interbellum* (Acco, 2009).

25. Eyal Weizman, *Forensic Architecture: Violence at the Threshold of Detectability* (Zone, 2017). This legal limit to the resolution of satellite imagery differs slightly from country to country, but typically sits at around 0.5–1 meter per pixel, roughly the space taken up by a human body when viewed from above. See also Houser, *Infowhelm*, 207.

26. On salvage accumulation, see Anna Tsing, *The Mushroom at the End of the World: On the Possibility of Life in Capitalist Ruins* (Princeton University Press, 2015).

27. Eduardo Gudynas, *Extractivisms, Politics, Economy and Ecology* (Columbia University Press, 2021), 6.

28. Axinocapital—Der Spezialist für Kapitalmehrung, "When is AVZ's Manono project eventually recognized as the largest lithium deposit in the world," YouTube, July 31, 2019, 7:42, https://www.youtube.com/watch?v=sBv7UExjk_A. At time of writing, there is considerable uncertainty around which companies will be involved in exploiting lithium in Monomo, with Chinese mining giant Zijin, the Congolese state-owned company Cominière, and more recently, the US company KoBold potentially involved. Attempts have been made on several occasions to encourage the processing of copper and cobalt within the DRC through

an export ban on concentrates, although these initiatives have often been reversed or unenforced. See Bobby Bascomb, "US Firm KoBold Metals Buys Stake in Contested Manono Lithium Project, DRC," *Mongabay*, May 28, 2025, https://news.mongabay.com/short-article/us-firm-kobold-metals-buys-stake-in-contested-manono-lithium-project-drc.

29. Fernando Henrique Cardoso and Enzo Faletto, *Dependency and Development in Latin America* (University of California Press, 1979).

30. For example, Senga brought 1 kilogram of sand from Manono to Brussels to contribute to a work by Marjolijn Dijkman in the *Charging Myths* exhibition. In a piece titled *Cloud to Ground #1* (2021–2022), Dijkman created artificial fulgurites by passing an electric current through the soil.

31. The exhibition subsequently toured to Framer Unframed in Amsterdam from February 23 to June 4, 2023.

32. As Tom Özden-Schilling has shown, the technoscientific labor of mineral prospecting has itself been thoroughly virtualized in recent decades, and since the early 2000s, some has been conducted by so-called "desktop prospectors" in countries such as Canada searching through public geoscience databases from their own homes. See Tom Özden-Schilling, "Desktop Prospecting and Extractivism at Home," *Environmental Humanities* 15, no. 3 (2023): 203–218.

33. Mike Fortun, *Promising Genomics: Iceland and deCODE Genetics in a World of Speculation* (University of California Press, 2008); Anna Lowenhaupt Tsing, *Friction: An Ethnography of Global Connection* (Princeton University Press, 2004). On biotechnology, see also Kaushik Sunder Rajan, *Biocapital: The Constitution of Postgenomic Life* (Duke University Press, 2006). And on expectations in STS more broadly, see Mads Borup et al., "The Sociology of Expectations in Science and Technology," *Technology Analysis and Strategic Management* 18, nos. 3–4 (2006): 285–298.

34. Tsing, *Friction*, 68.

35. Keith F. Davis, "Timothy O'Sullivan: Photographer," in *Timothy O'Sullivan: The King Survey Photographs*, ed. Keith F. Davis and Jane L. Aspinwall (Yale University Press, 2011), 75. As Davis points out, there were only two images of Native Americans in the entire body of photographs produced by O'Sullivan on these surveys.

36. As Arboleda has shown, new technologies in the mining industry used today can create "highly accurate representations of the subsurface, making the extraction of low-grade ore bodies profitable for the first time in history." Very few of these adopt a straightforward vertical perspective, and even the misleadingly named "vertical cartographies" pioneered in the late nineteenth century by the likes of Orra White Hitchcock create cross sections of the sub-surface of the earth viewed from the side. See Martín Arboleda, *Planetary Mine: Territories of Extraction under Late Capitalism* (Verso, 2020), 4.

37. Daniel Zizumbo-Villarreal and Patricia Colunga-GarcíaMarín, "Origin of Agriculture and Plant Domestication in West Mesoamerica," *Genetic Resources and Crop Evolution* 57 (2010): 813–825.

38. Allen Van Deynze et al., "Nitrogen Fixation in a Landrace of Maize Is Supported by a Mucilage-Associated Diazotrophic Microbiota" *PLOS Biology* 16, no. 8 (2018): e2006352.

39. Martha Pskowski, "Indigenous Maize: Who Owns the Rights to Mexico's 'Wonder' Plant?" *Yale Environment 360*, July 16, 2019, https://e360.yale.edu/features/indigenous-maize-who-owns-the-rights-to-mexicos-wonder-plant.

40. Edith Morales, *Aerial Root*, 2020, 3:59.

41. Nicolas Mateo et al., "Bioprospecting," in *Encyclopedia of Biodiversity*, vol. 1, ed. Simon Levin (McAfee, 2001), 471.

42. Cori Hayden, *When Nature Goes Public: The Making and Unmaking of Bioprospecting in Mexico* (Princeton University Press, 2003), 24.

43. Leanne M. Fecteau, "The Ayahuasca Patent Revocation: Raising Questions About Current U.S. Patent Policy," *Boston College Third World Law Journal* 21, no. 1 (2001): 69–104; Center for International Environmental Law, "Campaign Update: Protecting Traditional Knowledge: Ayahuasca Patent Dispute," https://www.ciel.org/project-update/protecting-traditional-knowledge-ayahuasca, accessed July 1, 2024; Cormac Sheridan, "EPO Neem Patent Revocation Revives Biopiracy Debate," *Nature Biotechnology* 23, no. 5 (2005): 511–513.

44. See, for example, Vandana Shiva, *Biopiracy: The Plunder of Nature and Knowledge* (South End, 1997); and numerous other publications by the same author.

45. *Nagoya Protocol on Access to Genetic Resources and the Fair and Equitable Sharing of Benefits Arising From Their Utilization to the Convention on Biological Diversity* (Secretariat of the Convention on Biological Diversity, 2011). Although adopted in 2010, the protocol only entered into force in 2014.

46. Hayden, *When Nature Goes Public*, 2–3.

47. Amy Quinton, "Study Finds Indigenous Mexican Variety of Corn Captures the Nitrogen It Needs from the Air," UC Davis Department of Plant Sciences, news release, August 7, 2018, https://www.plantsciences.ucdavis.edu/news/study-finds-indigenous-mexican-variety-corn-captures-nitrogen-it-needs-air.

48. In some cases, benefits have been shared with multiple communities who act as joint holders of traditional knowledge, such as the Khoikhoi and the San people of South Africa, who have a benefit-sharing agreement for the commercial use of rooibos. See Hayden, *When Nature Goes Public*, 2–3.

49. Noel Castree, "Bioprospecting: From Theory to Practice (and Back Again)," *Transactions of the Institute of British Geographers* 28, no. 1 (2003): 46.

50. For a critical reading of UPOV '91, see GRAIN, *UPOV 91 and Other Seed Laws: A Basic Primer on How Companies Intend to Control and Monopolise Seeds*, October 2015, 7.

51. Van Deynze et al., "Nitrogen Fixation in a Landrace of Maize."

52. Michel Foucault, *The Order of Things: An Archaeology of the Human Sciences* (Routledge, 1989); and Lorraine Daston and Katherine Pak, *Wonders and the Order of Nature, 1150–1750* (Zone, 2001).

53. Davis, "Timothy O'Sullivan," 87.

54. Daniela Bleichmar, "Books, Bodies, and Fields: Sixteenth-Century Transatlantic Encounters with the New World *Materia Medica*," in *Colonial Botany: Science, Commerce, and Politics in the Early Modern World*, ed. Londa Schiebinger and Claudia Swan (University of Pennsylvania Press, 2005), 99.

55. Kapil Raj, "Surgeons, Fakirs, Merchants, and Craftspeople: Making L'Empereur's *Jardin* in Early Modern South Asia," in *Colonial Botany*, 261.

56. *Yoo Cua Nuniri / Autosuficiencia*, curated by Oliver Martínez Kandt, Parallel Oaxaca, October 26–December 10, 2019. Another artwork that catalogues seeds in a similar fashion is Amar Kanwar's *272 Varieties of Indigenous, Organic Rice Seeds* (n.d.), in which the artist displays rice varieties from the Indian state of Odisha in handmade containers.

57. Edith Morales, interview with the author, February 19, 2021.

58. As Anke Te Hansen has shown, there is an old connection between botany and accountancy and "mercantile and scholarly bookkeeping techniques" evident in colonial botany from the early stages of its development. See Anke Te Hansen, "Accounting for the Natural World: Double-Entry Bookkeeping in the Field," in *Colonial Botany*, 239.

59. Morales, interview with the author, February 19, 2021.

60. For a discussion of how representations of volume come to matter in surface mining during European colonialization, current plans to mine the deep sea, and asteroid mining in outer space, see Yolanda Ariadne Collins et al., "'Mine the Volume': Excess and the Voluminous Ecological Politics of Capitalist Frontiers," *Environment and Planning E: Nature and Space* 8, no. 3 (2025): 955–977.

CHAPTER 3

1. Used with permission of John Wiley & Sons, from Howard T. Odum, *Environment, Power, and Society* (1971), 11; permission conveyed through Copyright Clearance Center, Inc.

2. I. Dundas et al., "Toxic Algal Bloom in Scandinavian Waters, May–June 1988," *Oceanography* 2, no. 1 (1989): 9–14.

3. H. Leivestad and B. Serigstad, "Some Observations on the Effects of *Chrysochromulina polylepis* on the Osmoregulation in Fish," *Norwegian Institute of Marine Research*, Rep. BKO 8803 (1988).

4. Dundas et al., "Toxic Algal Bloom in Scandinavian Waters," 14.

5. Roberta C. Hamme et al., "Volcanic Ash Fuels Anomalous Plankton Bloom in Subarctic Northeast Pacific," *Geophysical Research Letters* 37, no. 19 (2010): 1–5; Timothy R. Parsons and Frank A. Whitney, "Did Volcanic Ash from Mt. Kasatoshi in 2008 Contribute to a Phenomenal Increase in Fraser River Sock-Eye Salmon (*Oncorhynchus nerka*) in 2010?" *Fisheries Oceanography* 21, no. 5 (2012): 374–377.

6. Martin Lukacs, "World's Biggest Geoengineering Experiment 'Violates' UN Rules," *Guardian*, October 15, 2012, https://www.theguardian.com/environment/2012/oct/15/pacific-iron-fertilisation-geoengineering; Jeff Tollefson, "Ocean-Fertilization Project Off Canada Sparks Furore," *Nature* 490 (2012): 458–459.
7. The HSRC terminated George's contract in 2013. What followed was a series of lawsuits involving the HSRC and George, which resulted in the latter having to pay security for legal costs. *Ocean Pastures Corporation v. Old Masset Economic Development Corporation*, 2016 BCCA 12, https://www.bccourts.ca/jdb-txt/ca/16/00/2016BCCA0012.htm.
8. Michelle Allsopp et al., *A Scientific Critique of Oceanic Iron Fertilization as a Climate Change Mitigation Strategy* (Greenpeace Research Laboratories, 2007); Geoengineering Monitor, *Geoengineering Technology Briefing: Ocean Fertilization*, January 2021, https://www.geoengineeringmonitor.org/wp-content/uploads/2021/04/ocean-fertilization.pdf.
9. Tollefson, "Ocean-Fertilization Project Off Canada Sparks Furore." Although George sold himself to the Old Massett Village as "the world's leading expert" on ocean iron fertilization, he lacks many of the traditional credentials of expertise, such as academic qualifications. See Zach Horton, "Going Rogue or Becoming Salmon," *Cultural Critique* 97 (2017): 128–166; Kate Elizabeth Gannon and Mike Hulme, "Geoengineering at the 'Edge of the World': Exploring Perceptions of Ocean Fertilisation Through the Haida Salmon Restoration Corporation," *Geo: Geography and Environment* 5, no. 1 (2018): 1–21; Kate Elizabeth Gannon, "'40 Million Salmon Might Be Wrong': Ecological Worldviews and Geoengineering Technologies: The Case of the Haida Salmon Restoration Corporation" (PhD diss., King's College London, 2015).
10. David Keith, "Engineering the Planet," in *Climate Change Science and Policy*, ed. Stephen Schneider et al. (Island, 2010), 495.
11. John Shepherd, *Geoengineering the Climate: Science, Governance and Uncertainty* (Royal Society, 2009), ix.
12. See, for example, *39 Ways to Save the Planet*, produced by BBC and Royal Geographical Society, first aired December 21, 2020, on BBC Radio 4; *Can Geoengineering Save the Planet from Climate Change?*, produced by CBC, first aired October 11, 2017, https://www.cbc.ca/player/play/video/1.4351136. For an analysis of the US media coverage of Stratospheric Aerosol Injections specifically, see Toby Bolsen et al., "Public Response to Solar Geoengineering: How Media Frames About Stratospheric Aerosol Injection Affect Opinions," *Climatic Change* 176 (2023): article 112.
13. Clive Hamilton, *Earthmasters: The Dawn of the Age of Climate Engineering* (Yale University Press, 2013). One of the most prominent critics of geoengineering in recent years is Naomi Klein. See Klein, *This Changes Everything: Capitalism vs. the Climate* (Simon and Schuster, 2014).
14. For a book that does this work in an accessible fashion, see Holly Jean Buck, *After Geoengineering: Climate Tragedy, Repair and Restoration* (Verso, 2019).

15. There are numerous variations on the group's name. "Ocean Earth Construction and Development Corporation," "Ocean Earth Development Corporation," "Ocean Earth," and the acronym "OEDC" all appear in the group's communications, and sometimes more than one name is used in the same publication. As well as Fend, some of the early contributors to Ocean Earth were Colen Fitzgibbon, Jenny Holzer, Peter Nadin, Richard Prince, and Robin Winters, although countless others have been involved since.

16. David Joselit and Rachel Harrison, "A Conversation with Peter Fend," *October* 125 (2008): 117–136.

17. Paul N. Edwards, *A Vast Machine: Computer Models, Climate Data, and the Politics of Global Warming* (MIT Press, 2010).

18. Shepherd, *Geoengineering the Climate*, v.

19. Shepherd, *Geoengineering the Climate*, 49.

20. Silke Beck and Martin Mahony, "The Politics of Anticipation: The IPCC and the Negative Emissions Technologies Experience," *Global Sustainability* 1, no. e8 (2018): 1–8; IPCC Core Writing Team et al., eds., *Climate Change 2014: Synthesis Report. Contribution of Working Groups I, II and III to the Fifth Assessment Report of the Intergovernmental Panel on Climate Change* (IPCC, 2014), 89. More recent working group reports from the IPCC also explore solar radiation management and CO_2 removal in detail, incorporating the latter into its emissions pathways. See Priyadarshi R. Shukla et al., eds., *Climate Change 2022: Mitigation of Climate Change. Contribution of Working Group III to the Sixth Assessment Report of the Intergovernmental Panel on Climate Change* (IPCC, 2022).

21. For a project overview, see UKRI, "UK Invests Over £30m in Large-Scale Greenhouse Gas Removal," news release, May 24, 2021, https://co2re.org/uk-invests-over-30m-in-large-scale-greenhouse-gas-removal.

22. Jeff Tollefson, "First Sun-Dimming Experiment Will Test a Way to Cool Earth," *Nature* 563 (2018): 613–615. The project is called the Stratospheric Controlled Perturbation Experiment (SCoPEx), and is one of several geoengineering research projects to be funded by Bill Gates. For more details, see "SCoPEx: Stratospheric Controlled Perturbation Experiment," Keutsch Group at Harvard, https://www.keutschgroup.com/scopex#h.p_Xru29Emo-OMw, accessed June 1, 2021. For the letter regarding SCoPEx from the Saami council, see Christina Henriksen et al., open letter to Members of the SCoPEx Advisory Committee, February 24, 2021, https://www.saamicouncil.net/news-archive/open-letter-requesting-cancellation-of-plans-for-geoengineering.

23. Nicolás Rivero, "Could spraying Sea Salt into the Clouds Cool the Planet?" *Washington Post*, June 4, 2024, https://www.washingtonpost.com/climate-solutions/2024/06/04/marine-cloud-brightening-geoengineering-alameda. A study based on modeling the effects of marine cloud brightening off the coast of California suggests that the technique could become less effective over time and even contribute to increasingly frequent heatwaves in Europe. See Jessica S. Wan et al., "Diminished Efficacy of Regional Marine Cloud Brightening in a Warmer

World," *Nature Climate Change* 14 (2024): 808–814. At the time of writing, significant funding has become available for research into SRM in the UK, with the UK Government's Advanced Research and Invention Agency (ARIA) providing 58.8 million pounds for a project that incorporates fields trials for stratospheric sulfur injection. See "Exploring Climate Cooling," ARIA, https://www.aria.org.uk/opportunity-spaces/future-proofing-our-climate-and-weather/exploring-climate-cooling, accessed July 1, 2025. See also Daniel Patrick Harrison, "Could 'Marine Cloud Brightening' Reduce Coral Bleaching on the Great Barrier Reef?" *Conversation*, October 16, 2023, https://theconversation.com/could-marine-cloud-brightening-reduce-coral-bleaching-on-the-great-barrier-reef-214308.

24. Paul J. Crutzen, "Albedo Enhancement by Stratospheric Sulfur Injections: A Contribution to Resolve a Policy Dilemma?" *Climate Change* 77, no. 3 (2006): 216.

25. Jack Stilgoe, *Experiment Earth: Responsible Innovation in Geoengineering* (Routledge, 2015), 93.

26. James Rodger Fleming, *Fixing the Sky: The Checkered History of Weather and Climate Control* (Columbia University Press, 2010), 236. The proposals were made in a paper by M. Ye. Shvets, "Control and Climate Through Stratospheric Dusting," Library of Congress, Aerospace Information Division, "Climate Control"; see also John Von Neumann, "Can We Survive Technology?" *Fortune Magazine* 6 (June 1955): 108.

27. Ben Bova, *The Weathermakers* (Holt, Rinehart and Winston, 1967); Leonard Leokum and Paul Posnick, *Weather War* (Arrow, 1979); J. G. Ballard, *Vermillion Sands* (Vintage Classics, 2001).

28. For example, a report produced for the President's Scientific Advisory committee in 1965 included an appendix on carbon dioxide and global warming, but was ostensibly concerned with pollution. John W. Tukey et al., *Restoring the Quality of Our Environment: Report of the Environmental Pollution Panel, President's Science Advisory Committee* (Government Printing Office, 1965).

29. Charles D. Keeling, "Rewards and Penalties of Monitoring the Earth," *Annual Review of Energy and the Environment* 23 (1998): 25–82. As mentioned in chapter 1, other early publications on atmospheric CO2 were produced by Gilbert N. Plass. For discussion of this and the more general development of climate science, see James Rodger Fleming, *Historical Perspectives on Climate Change* (Oxford University Press, 1998); and Spencer Weart, *The Discovery of Global Warming* (Harvard University Press, 2004).

30. Fleming, *Fixing the Sky*. See also Chunglin Kwa, "The Rise and Fall of Weather Modification: Changes in American Attitudes Towards Technology, Nature, And Society," in *Changing the Climate: Expert Knowledge and Environmental Governance*, ed. Paul Edwards and Clark Miller (MIT Press, 2001). Of all regions, it is perhaps the United Arab Emirates and China that use cloud seeding the most today. See Alya Al Mazroui and Sufian Farrah, "The UAE Seeks Leading Position in Global Rain Enhancement Research," *Journal of Weather Modification*, 49, no. 1 (2017): 54–59; and Josh Ye, "China Showers 1.15 Billion Yuan on Rainmaking Project for Parched Northwest," *South China Morning Post*, January 24, 2017,

https://www.scmp.com/news/china/policies-politics/article/2064729/china-showers-y115b-rainmaking-project-parched.

31. Fleming, *Fixing the Sky*.

32. Hamilton, *Earthmasters*. See also Robert Lempert and Don Prosnitz, *Governing Geoengineering Research: A Political and Technical Vulnerability Analysis of Potential Near-Term Options*, RAND Report No. TR-846-RC (RAND Corporation, 2011); and Joseph O. Fletcher, *Managing Climate Resources*, RAND Report No. P-4000-1 (RAND Corporation, 1969).

33. Kathryn Yusoff, "The Geoengine: Geoengineering and the Geopolitics of Planetary Modification," *Environment and Planning A* 45, no. 12 (2013): 2802.

34. One of many other patent applications to have arisen from the commercially oriented US academic system is from Leslie A. Field, a professor based at Stanford University, who lays claim to a series of "Methods for cooling water temperature using high albedo materials." This involves scattering tiny balls ranging from 100 to 3 millimeters in diameter in order to increase the reflectivity of the water's surface (patent document identifier US 20200288651 A1). Despite having authored a blog post about the problems of asserting intellectual property over geoengineering technologies, David Keith also holds a number of patents in the field. See David Keith and John Dykema, "Why We Chose Not to Patent Solar Geoengineering Technologies," *David Keith* (blog) May 3, 2018, https://davidkeith.earth/why-we-chose-not-to-patent-solar-geoengineering-technologies. See also Geoengineering Monitor, *Ocean Fertilization*, 2.

35. Inhabitants, "A Brief History of Geoengineering," September 2015, video, 6:33, archived at https://web.archive.org/web/20190730155008/https://inhabitants-tv.org/sept2015_geoengineering.html.

36. Paul N. Edwards, *A Vast Machine: Computer Models, Climate Data, and the Politics of Global Warming* (MIT Press, 2010).

37. Cesare Marchetti, "On Geoengineering and the CO_2 Problem," *Climate Change* 1 (1977): 67. As Stilgoe points out, the term "geoengineering" predates this publication by at least fifteen years. An early use can be found in L. V. Berkner, "Geoscience and Geoengineering," *Proceedings of the IRE* 50, no. 11 (1962): 2180–2182, albeit not in the context of climate change research. Here, "geoengineering" is used as a catch-all term for "applied geoscience" (2182).

38. "Systems analysis," as one of several approaches to strategic thinking incubated within Cold War organizations such as RAND, was developed by Edward Paxton in the late 1940s and shared a common root in Operations Research (OR), which involved viewing war through the prism of rational (typically mathematical) analysis. See Paul N. Edwards, *The Closed World: Computers and the Politics of Discourse in Cold War America* (MIT Press, 1997).

39. Dian R. Hitchcock and James E. Lovelock, "Life Detection by Atmospheric Analysis," *Icarus* 7, no. 2 (1967): 149–159; James E. Lovelock, "Gaia as Seen Through the Atmosphere," *Atmospheric Environment* 6 (1972): 579–580; James E. Lovelock and Lynn Margulis, "Atmospheric Homeostasis by and for the Biosphere: The

Gaia Hypothesis," *Tellus* 26 (1974): 2–10; James E. Lovelock and Lynn Margulis, "Homeostatic Tendencies of the Earth's Atmosphere," *Origins of Life* 5 (1974): 93–103; Lynn Margulis and James E. Lovelock, "Biological Modulation of the Earth's Atmosphere," *Icarus* 21 (1974): 471–489.

40. As Margulis stated later, "No organism eats its own waste." Lynn Margulis, "Gaia Is a Tough Bitch," in *The Third Culture: Beyond the Scientific Revolution*, ed. John Brockman (Simon & Schuster, 1995), 140. More recently, Bruno Latour and Isabelle Stengers have sought to extricate Gaia from its roots in cybernetics and systems theory. See Bruno Latour, "Why Gaia Is Not a God of Totality," *Theory, Culture & Society* 34, no. 2–3 (2017): 61–81; and Isabelle Stengers, *In Catastrophic Times: Resisting the Coming Barbarism* (Open Humanities, 2015).

41. James E. Lovelock, "A Physical Basis for Life Detection Experiments," *Nature* 207, no. 4997 (1965): 568–570; Timothy M. Lenton et al., "Gaia as Seen from Within," *Theory, Culture & Society* 41, no. 5 (2024): 69–90.

42. On the bottom up, see, for example, James E. Lovelock, "Natural Halocarbons in the Air and in the Sea," *Nature* 256, no. 5514 (1975): 193–194.

43. Erik M. Conway, "Bringing NASA Back to Earth: A Search for Relevance During the Cold War," in *Science and Technology in the Global Cold War*, ed. Naomi Oreskes and John Krige (MIT Press, 2014), 261.

44. Odum, *Environment, Power, and Society*, 9 and 11.

45. Odum, *Environment, Power, and Society*, 10–11.

46. Eugene P. Odum and Gray W. Barrett, *Fundamentals of Ecology*, 5th ed. (Thomson Brooks/Cole, 2005), 69.

47. Joel Hagen, "Ecology and the Atomic Age," in *An Entangled Bank: The Origins of Ecosystem Ecology* (Rutgers University Press, 1992).

48. Howard T. Odum, *Systems Ecology: An Introduction* (Wiley, 1983), 579. For a detailed discussion of the Odums, see M. G. Madison, "'Potatoes Made of Oil': Eugene and Howard Odum and the Origins and Limits of American Agroecology," *Environment and History* 3, no. 2 (1997): 209–238.

49. Odum, *Environment, Power, and Society*, 279.

50. Haraway, "Situated Knowledges," 582.

51. Yusoff, "Geoengine," 2806. A similar argument is made in Maia Galarraga and Bronislaw Szerszynski, "Making Climates: Solar Radiation Management and the Ethics of Fabrication," in *Engineering the Climate: The Ethics of Solar Radiation Management*, ed. Christopher Preston (Lexington Books, 2012). Complementing these onto-epistemological discussions of geoengineering is writing by T. J. Demos, who speaks directly of the role of visual culture, and the way "imaging systems . . . tend to grant viewers a sense of control over the represented object of their gaze, even if that control is far from reality." See Demos, *Against the Anthropocene*, 28. See also T. J. Demos, "To Save a World: Geoengineering, Conflictual Futurisms, and the Unthinkable," *e-flux* 94 (October 2018), https://www.e-flux.com/journal/94/221148/to-save-a-world-geoengineering-conflictual-futurisms-and-the-unthinkable.

52. For an extended discussion of the concept, see Elgė Rindzevičiūtė, "Soviet Policy Sciences And Earth System Governmentality," *Modern Intellectual History* 17, no. 1 (2020): 179–208. It is notable that, despite Cold War tensions, Vernadsky makes an appearance in the Odums's textbook *Fundamentals of Ecology*.

53. As one example of his work on global climate modeling, Moiseyev was a key contributor to research on a possible "nuclear winter" that modeled the effect that large quantities of soot and smoke created by nuclear conflict would have on the climate.

54. Eglė Rindzevičiūtė, *The Will to Predict: Orchestrating the Future Through Science* (Cornell University Press, 2023).

55. Rindzevičiūtė, *Will to Predict*, 171–172.

56. James E. Lovelock and Chris G. Rapley, "Ocean Pipes Could Help the Earth to Cure Itself," *Nature* 449, no. 7161 (2007): 403; James Lovelock, *The Vanishing Face of Gaia: A Final Warning: Enjoy It While You Can* (Penguin, 2010).

57. Joselit and Harrison, "Conversation with Peter Fend," 131.

58. Ocean Earth, "Satellite Monitoring of the Earth Is Architectural Endeavour [1987]," in *Ocean Earth: 1980 Bis Heute*, ed. Peter Weiber et al. (Neue Galerie am Landesmuseum Joanneum, 1994), 53.

59. These blind spots became even more apparent when considering Fend's proposal to use water as a principle to reorganize Afghanistan and Iran, shown to me in an interview. Peter Fend, interview with the author, July 20, 2021.

60. Fend makes direct reference to Physiocracy in, among other places, a short piece in The Buckminster Fuller Challenge Prize. See Peter Fend, "Eco-Tax," Buckminster Fuller Institute, 2015, https://www.bfi.org/ideaindex/projects/2015/eco-tax.

61. Joselit and Harrison, "Conversation with Peter Fend," 131; Ocean Earth, "Agriculture Ends, Art Takes Over [1976]," in *Ocean Earth: 1980 Bis Heute*, ed. Peter Weiber et al. (Neue Galerie am Landesmuseum Joanneum, 1994), 6.

62. Fend, interview with the author, July 20, 2021.

63. Joselit and Harrison, "Conversation with Peter Fend," 132.

64. For further reading on ecofascism, see Michael E. Zimmerman, "Ecofascism: An Enduring Temptation," in *Environmental Philosophy: From Animal Rights to Radical Ecology*, 4th ed., ed. Micahel E. Zimmerman et al. (Prentice Hall, 2004); and The Zetkin Collective and Andreas Malm, *White Skin, Black Fuel: On the Dangers of Fossil Fascism* (Verso, 2021).

65. Hamilton, *Earthmasters*.

66. Weibel et al., *Ocean Earth*, 22.

67. Fend, interview with the author, July 20, 2021.

68. Weibel et al., *Ocean Earth*, 93. While not technically a fossil fuel when produced in this way, methane nevertheless produces significant amounts of CO_2 when burnt.

69. Fend, interview with the author, July 20, 2021. One of the examples Fend cites is a project sponsored by the Tokyo Metropolitan Government, who prohibited him from making a fully functioning seaweed rig on the grounds that this would require an act of legislation and rigorous evaluation procedure before it could be used. Instead, a maquette was constructed at a cost Fend estimates was higher than the real thing would have been.

70. In *Ocean Earth*, Fend pays direct homage to the Ocean Food and Energy Farm Project, featuring an illustration by the project leader Howard A. Wilcox alongside his own concept sketches. See also Donald F. Squires, *Alchemy for the '80s: Riches from Our Coastal Resources* (New York Sea Grant Institute, 1982), for a similar set of proposals from the time.

71. Holly Jean Buck, *After Geoengineering: Climate Tragedy, Repair and Restoration* (Verso, 2019), 80; Zi-Min Hu et al., "Kelp Aquaculture in China: A Retrospective and Future Prospects," *Reviews in Aquaculture* 13 (2021): 1324–1351.

72. Dorte Krause-Jensen and Carlos Duarte, "Substantial Role of Macroalgae in Marine Carbon Sequestration," *Nature Geoscience* 9 (2016): 737–742; Dorte Krause-Jensen et al., "Sequestration of Macroalgal Carbon: The Elephant in the Blue Carbon Room," *Biological Letters* 14, no. 6 (2018): 1–6.

73. In many ways, Holly Jean Buck tries to overcome this dichotomy, articulating selective support for geoengineering from a broadly left perspective, broaching issues of climate governance, and emphasizing community-run negative emissions technologies that would demand new forms of political structure.

74. Fend, interview with the author, July 20, 2021.

75. Bradley Cantrell et al., "Designing Autonomy: Opportunities for New Wildness in the Anthropocene," *Trends in Ecology and Evolution* 32, no. 3 (2017): 162.

76. Anne E. Thessen, "Adoption of Machine Learning Techniques in Ecology and Earth Science," *One Ecosystem* 1 (2016): 1–38.

77. Isha Salian, "Rock On: Scientists Use AI to Improve Sequestering Carbon Underground," Nvidia (blog), April 8, 2022, https://blogs.nvidia.com/blog/ai-improves-carbon-sequestration; Ben Kravitz et al., "Comparing Different Generations of Idealized Solar Geoengineering Simulations in the Geoengineering Model Intercomparison Project (GeoMIP)," *Atmospheric Chemistry and Physics* 21, no. 6 (2021): 4231–4247. More details of the work carried out as part of the Geoengineering Model Intercomparison Project can be found at https://climate.envsci.rutgers.edu/GeoMIP/index.html. See also Anthony C. Jones et al., "Impacts of Hemispheric Solar Geoengineering on Tropical Cyclone Frequency," *Nature Communications* 8, no. 1 (2017): 1382.

78. Daniele Visioni et al., "Is Turning Down the Sun a Good Proxy for Stratospheric Sulfate Geoengineering?" *Journal of Geophysical Research: Atmospheres* 126 (2021): e2020JD033952. For a critique of "the socio-technical imaginary of the thermostat," see Mike Hulme, *Can Science Fix Climate Change? A Case Against Climate Engineering* (Polity, 2014), 82.

79. Tega Brain, "The Environment Is Not a System," *A Peer-Reviewed Journal About* 7, no. 1 (2018): 156.

80. I have written elsewhere on how environments are rendered calculable through such financial instruments as biodiversity offsets. See Josh Bowsher and Theo Reeves-Evison, "On Capital's Watch: Derivative Ecology and the Temporal Logic of Biodiversity Credits," *New Formations* 99 (2019): 33–51.

81. Tega Brain, interview with the author, February 24, 2021.

82. Artists' description of Asunder, available from official project website: https://asunder.earth.

83. Jussi Parikka, *Operational Images: From the Visual to the Invisual* (University of Minnesota Press, 2023), 141.

84. Brain, "Environment Is Not a System," 153.

85. Dan Brockington et al., *Nature Unbound: Conservation, Capitalism and the Future of Protected Areas* (Earthscan, 2008), 198.

86. Brain, interview with the author, February 24, 2021.

87. For an artistic exploration of AI's capacity to hallucinate, see *Trevor Paglen: Adversarially Evolved Hallucinations*, ed. Anthony Downey (Sternberg, 2024).

88. For another artistic project that interrogates planetary visual cultures and the infrastructures of scientific imaging, see the work of Geocinema (Asia Bazdyrieva and Solveig Qu Suess), in particular their film/exhibition *Making of Earths* (2021).

89. Julian Oliver et al., "The Critical Engineering Manifesto," 2021, https://criticalengineering.org.

90. The study of animal-computer interactions (ACI) is one recent field of study where more-than-humans are treated as users. See Sarah E. Ritvo and Robert S. Allison, "Designing for the Exceptional User: Nonhuman Animal-Computer Interaction," *Computers in Human Behavior* 70 (2017): 222–233.

91. Carl Folke et al., "Catastrophic Shifts in Ecosystems," *Nature* 413, no. 6856 (2002): 591–596; Marten Scheffer and Stephen R. Carpenter, "Catastrophic Regime Shifts in Ecosystems: Linking Theory to Observation," *Trends in Ecology and Evolution* 18, no. 12 (2003): 648–656.

92. Brain, interview with the author, February 24, 2021. See also Tega Brain, "Eccentric Engineering: Thoughts for the Anthropocene," YouTube, July 20, 2015, 15:29, https://www.youtube.com/watch?v=YhJR_iN6QBM.

93. Shoshana Zuboff, *The Age of Surveillance Capitalism: The Fight for a Human Future at the New Frontier of Power* (Profile, 2019).

94. Brain, interview with the author, February 24, 2021.

95. The centralized political planning presupposed by many geoengineering proposals could be aligned to what Andreas Malm calls "geoeconomies." See Andreas Malm, "Planning the Planet: Geoengineering Our Way Out of and Back into a Planned Economy," in *Has It Come to This: The Promises and Perils of Geoengineering on the Brink*, ed. J. P. Sapinski et al. (Rutgers University Press, 2020).

96. Céline Granjou et al., "The Politics of Anticipation: On Knowing and Governing Environmental Futures," *Futures* 92 (2017): 5–11.

97. Neil Smith, "There's No Such Thing as a Natural Disaster," *Items: Insights from the Social Sciences*, June 11, 2006, https://items.ssrc.org/understanding-katrina/theres-no-such-thing-as-a-natural-disaster.

98. Sushil Kumar Dash et al., "Some Evidence of Climate Change in Twentieth-Century India," *Climatic Change* 85 (2007): 299–321; Rishi Sahastrabuddhe et al., "Indian Summer Monsoon Rainfall in a Changing Climate: A Review," *Journal of Water and Climate Change* 14, no. 4 (2023): 1061–1088; Amita Kumari and Alok Kumar Mishra, "How Is Climate Change Altering the Precipitation Pattern Over Northwestern India?" *International Journal of Climatology* 44, no. 8 (2024): 2592–2606.

99. To cite two examples, on August 29, 2017, a flood in Mumbai caused more devastation, cutting off power and disrupting rail and road infrastructure. In June 2021, more disruption was caused by a much smaller quantity of rain, 60 millimeters in an hour. This has prompted a public debate on why the nullahs haven't been cleared as intended, and who is responsible for the failure.

100. Anuradha Mathur and Dilip da Cunha, *SOAK: Mumbai in an Estuary* (Rupa, 2009), 46.

101. Mathur and da Cunha, *SOAK*, 39.

102. Dilip da Cunha, *The Invention of Rivers: Alexander's Eye and Ganga's Descent* (University Pennsylvania Press: 2019), 290.

103. Mathur and da Cunha, *SOAK*, 71.

104. Matthew H. Edney, "Defining a Unique City: Surveying and Mapping Bombay after 1800," in *Bombay to Mumbai: Changing Perspectives*, ed. Pauline Rohatgi et al. (Marg, 1997), 50. There are of course more recent cartographic practices that incorporate temporal elements. See, for example, Sybille Lammes et al., eds., *Time for Mapping: Cartographic Temporalities* (Manchester University Press, 2018).

105. Mathur and da Cunha, *SOAK*, 186. For an attempt to produce a dynamic model of Australia's coastlines that takes into account tidal variations, see Robbi Bishop-Taylor et al., "Between the Tides: Modelling the Elevation of Australia's Exposed Intertidal Zone at Continental Scale," *Estuarine, Coastal and Shelf Science* 223 (2019): 115–128.

106. Victoria Carchidi, "Review: *SOAK: Mumbai in an Estuary*," *Electronic Green Journal* 1, no. 31 (2011): article 7; on sponge cities, see Hui Li et al., "Sponge City Construction in China: A Survey of the Challenges and Opportunities," *Water* 9, no. 9 (2017): 594.

107. Mathur and da Cunha, *SOAK*, 186.

108. Mathur and da Cunha, *SOAK*, 46.

109. Steve Lerner, *Sacrifice Zones: The Front Lines of Toxic Chemical Exposure in the United States* (MIT Press, 2010); Ryan Juskus, "Sacrifice Zones: A Genealogy and

Analysis of an Environmental Justice Concept," *Environmental Humanities* 15, no. 1 (2023): 3–24.

110. On this, see Tsing, *Friction*. Tsing writes of the ways in which scale can be nuanced in climate models, but nevertheless still freighted with the assumption that "scales ought to fit neatly inside each other, the small inside the large, each neutral and fully encompassed by the next scale up. . . . The incompatibility of scales is a practical problem. It does not challenge the principles of the model, which requires that the local disappear compatibly inside the global" (104).

111. Tsing, *Friction*, 103.

112. Johann Feichter and Markus Quante, "From Predictive to Instructive: Using Models for Geoengineering," in *Culture of Prediction in Atmospheric and Climate Science: Epistemic and Cultural Shifts in Computer-Based Modelling and Simulation*, ed. Matthias Heymann et al. (Routledge, 2017), 187. For a compelling account of how climate change exceeds the limits of the knowable, particularly in the case of forest fires in North America, see Adriana Petryna, *Horizon Work: At the Edges of Knowledge in an Age of Climate Change* (Princeton University Press, 2022).

113. On the displacement of the human as the principle subject of history, see Dipesh Chakrabarty, *The Climate of History in a Planetary Age* (University of Chicago Press, 2021). Chakrabarty proposes that the global, as a category of historical experience that "refers to matters that happen within human horizons of time—the multiple horizons of existential, intergenerational and historical time" is increasingly being replaced by an awareness of what he calls the "planet," which operates according to a different set of time scales and shatters human illusions of control (86). Influential to this growing awareness, Chakrabarty argues, is the development of earth systems science as a means of apprehending the planet. From the perspective of this chapter, it is worth emphasizing that the kind of systems-based thinking discussed by Chakrabarty can in some instances prevent recognition of non-systemic features of planetary existence.

114. Lovelock, *Vanishing Face of Gaia*, 139.

115. In "Making Climates," Galarraga and Szerszynski associate the activity of master-planning in this way more with the (ideal) figure of the "climate architect," "who knows in advance the form that they want the climate to take; who can identify the process whereby they can provoke the climate to take it; and who can carry out that process and bring the matter of climate into the desired form" (228–229).

CHAPTER 4

1. Giorgos Kallis, *Degrowth* (Agenda, 2018), 138.

2. Giorgos Kallis and Hug March, "Imaginaries of Hope: The Utopianism of Degrowth," *Annals of the Association of American Geographers* 105, no. 2 (2015): 361. Copyright © 2013 by the American Association of Geographers. Reprinted by

permission of Informa UK Limited, trading Taylor & Francis Group, https://www.tandfonline.com, on behalf of American Association of Geographers.

3. For an example of mainstream politics embracing degrowth, see Alberto Garzón Espinosa, "The Limits to Growth: Eco-Socialism or Barbarism," *Monthly Review* 74, no. 3 (July–August 2022), https://monthlyreview.org/2022/07/01/the-limits-to-growth-ecosocialism-or-barbarism. At time the article was written, Garzón was general coordinator of United Left party in Spain, and Minister of Consumption Affairs of Spain's Government. In its 2022 report *Impacts, Adaptation and Vulnerability*, the IPCC mentions degrowth fifteen times, with additional mentions in subsequent reports. See Hans-Otto Pörtner et al., eds., Climate *Change 2022: Impacts, Adaptation and Vulnerability. Contribution of Working Group II to the Sixth Assessment Report of the Intergovernmental Panel on Climate Change* (IPCC, 2022).
4. François Schneider, quoted in "Recherche pour la décroissance," http://www.decroissance.org/francois/recherche/index.htm, accessed March 1, 2020.
5. As documented on Schneider's personal website. See François Schneider, "Marche pour la décroissance de toutes les tumeurs," http://www.decroissance.org/francoisv, accessed March 1, 2020.
6. Kallis and March, "Imaginaries of Hope," 361.
7. Michel Bosquet (A pen name of André Gorz), *Nouvel Observateur*, Paris, 397, June 19, 1972, proceedings from a public debate organized in Paris by the Club du Nouvel Observateur, iv; André Gorz, *Écologie et liberté* (Galilée, 1977); André Amar, "La croissance et le problème moral," *Cahiers de la Nef* 52 (1973): 133; Nicholas Georgescu-Roegen, *Demain la décroissance: Entropie-écologie-économie* (Pierre-Marcel Favre, 1979). Herman E. Daly, one of the founders of ecological economics, was also advocating for a "steady state economics" in the 1970s. See Herman E. Daly, *Towards a Steady State Economy* (W. H. Freeman, 1973).
8. The slogans used in these protests were *decrescita*, *decreixement*, and *decrecimiento* in Italian, Catalan, and Spanish, respectively.
9. For a critique of "folk politics," see Nick Srnicek and Alex Williams, *Inventing the Future: Postcapitalism and a World Without Work* (Verso, 2015). For two pointed critiques of degrowth, see Leigh Phillips, *Austerity Ecology and the Collapse Porn Addicts: A Defense of Growth, Progress, Industry and Stuff* (Zero, 2015); and Matt Huber, "Ecosocialism: Dystopian and Scientific," *Socialist Forum* (Winter 2019), https://sfarchive.dsausa.org/issues/winter-2019/ecosocialism-dystopian-and-scientific. A response to the latter article has been written by one of the most prominent advocates of degrowth. See Giorgos Kallis, "Degrowth Is Utopian, and That's a Good Thing," *Uneven Earth*, April 26, 2019, https://unevenearth.org/2019/04/degrowth-is-utopian-and-thats-a-good-thing.
10. See note 3 of this chapter, and for a discussion of the significance of the IPCC's mentioning of degrowth, see the blog of social scientist and economist Timothée Parrique, especially "Degrowth in the IPCC AR6 WGII," March 5, 2022, https://timotheeparrique.com/degrowth-in-the-ipcc-ar6-wgii; and "Degrowth in the IPCC AR6 WGIII," April 7, 2022, https://timotheeparrique.com/degrowth-in-the-ipcc-ar6-wgiii.

11. Johan Galtung, "On Future Research and Its Role in the World," in *Challenges from the Future: Proceedings of the International Future Research Conference*, vol. 1, compiled by the Japan Society of Futurology (Kodansha, 1970), 105.
12. Chris Carlsson, *Nowtopia: How Pirate Programmers, Outlaw Bicyclists, and Vacant-Lot Gardeners Are Inventing the Future Today* (AK, 2008).
13. Barbara Adam, *Timescapes of Modernity: The Environment & Invisible Hazards* (Routledge, 1998), 11.
14. Adam, *Timescapes of Modernity*, 9.
15. Adam, *Timescapes of Modernity*, 11.
16. Gareth Dale, "Seventeenth-Century Origins of the Growth Paradigm," in *History of the Future of Economic Growth*, ed. Iris Borowy and Matthias Schmelzer (Routledge, 2017), 46.
17. Dale, "Seventeenth-Century Origins," 34. This point is also made in Till Düppe, *The Making of the Economy: A Phenomenology of Economic Science* (Lexington, 2011): "Scrutinising 'the causes of growth,' as Adam Smith later would do, was already based on the neutralization of time, and in particular the cyclical time that dominated economic life before capitalism. This question would make no sense in a world that ran *cyclically* around some kind of transcendent order—be it the Greek cosmos or the Christian divine realm" (89).
18. Simon Kuznets, *National Income, 1929–1932*, 73rd US Congress, 2nd session, Senate document no. 124 (1934), 5–7. It was the 1944 Bretton Woods conference that established the measure of GDP as an international standard.
19. On the imposition of GDP targets on the so-called "third world," see Arturo Escobar, *Encountering Development: The Making and Unmaking of the Third World* (Princeton University Press, 1995).
20. William Davies, "Introduction to Economic Science Fictions," in *Economic Science Fictions*, ed. William Davies (Goldsmiths, 2018), 15.
21. Angus Maddison, *The World Economy*, vol. 1, *A Millennial Perspective* (OECD, 2001); and Angus Maddison, *The World Economy*, vol. 2, *Historical Statistics* (OECD, 2003).
22. Kallis, *Degrowth*, 9.
23. Serge Latouche, *Farewell to Growth* (Polity, 2009), 8.
24. To take two examples, during his time as prime minister of the UK, David Cameron proposed the creation of a "wellbeing index," and in 2019, the government of New Zealand, led by Jacinda Arden, unveiled a "wellbeing budget." See Michael Mintrom, "New Zealand's Wellbeing Budget Invests in Population Health," *Milbank Quarterly* 97, no. 4 (December 2019): 893–896. See also Will Davies, *The Happiness Industry: How the Government and Big Business Sold Us Well-Being* (Verso, 2015).
25. Adam, *Timescapes of Modernity*, 14.
26. Adam, *Timescapes of Modernity*, 76.
27. Adam, *Timescapes of Modernity*, 9.

28. See, for example, Sandra Calkins and Tyler Zoanni, "Growth," *Environmental Humanities* 16, no. 1 (2024): 265–269; and Sandra Calkins, "Health as Growth: Bananas, Humanitarian Biotech, and Human-Plant Histories in Uganda," *Medicine Anthropology Theory* 6, no. 2 (2019): 29–53.

29. Michelle Bastian, "Fatally Confused: Telling the Time in the Midst of Ecological Crises," *Environmental Philosophy* 9 (2012): 23–48; Michelle Bastian, "Encountering Leatherbacks in Multispecies Knots of Time," in *Extinction Studies: Stories of Time, Death and Generations*, ed. Thom van Doreen et al. (Columbia University Press, 2017); Deborah Bird Rose, "Multispecies Knots of Ethical Time," *Environmental Philosophy* 9, no. 1 (2012): 127–140. See also "Temporal Environments: Rethinking Time and Ecology," ed. Jacob Metcalf and Thom van Doreen, special issue, *Environmental Philosophy* 9, no. 1 (2012).

30. Rose, "Multispecies Knots of Ethical Time," 129–130.

31. Rose, "Multispecies Knots of Ethical Time," 138.

32. Ottar N Bjørnstad et al., "Nonlinearity and Chaos in Ecological Dynamics Revisited," *Proceedings of the National Academy of Sciences of the United States of America* 112, no. 20 (2015): 6252–6253; Chihhao Hsieh et al., "Extending Nonlinear Analysis to Short Ecological Time Series," *American Naturalist* 171, no. 1 (2008): 71–80; Jan M. Baert et al., "Nonlinear Partitioning of Biodiversity Effects on Ecosystem Functioning," *Methods in Ecology and Evolution* 8, no. 10 (2017): 1233–1240.

33. Frederic Clements, *Plant Succession: An Analysis of the Development of Vegetation* (Carnegie Institution of Washington, 1916); Carl Folke et al., "Catastrophic Shifts in Ecosystems," *Nature* 413, no. 6856 (2002): 591–596; Marten Scheffer and Stephen R. Carpenter, "Catastrophic Regime Shifts in Ecosystems: Linking Theory to Observation," *Trends in Ecology and Evolution* 18, no. 12 (2003): 648–656.

34. Together with Josh Bowsher, I've written about these nonlinear dynamics in the context of attempts to quantify and offset biodiversity losses. See Josh Bowsher and Theo Reeves-Evison, "On Capital's Watch: Derivative Ecology and the Temporal Logic of Biodiversity Credits," *New Formations* 99 (2019): 33–51.

35. T. J. Clark and Angela D. Luis, "Nonlinear Population Dynamics Are Ubiquitous in Animals," *Nature Ecology and Evolution* 4 (2020): 75–81.

36. Growing degree days represent an adaptation of clock time to the patterns of development in the life cycles of plants and insects, each having a minimum and maximum temperature within which growth will occur. As such, they represent a way of measuring "thermal time."

37. Troy Conrad Therrien, "Along for the Ride: Curator's Introduction," in *Countryside, A Report*, ed. Julius Wiedemann (Taschen, 2020), 14.

38. Felicity D. Scott, "Futures Market," *Artforum* 58, no. 10 (July/August 2020): 126.

39. Therrien, "Along for the Ride," 12–13.

40. In recent years the Netherlands has consistently ranked in the top three in global food exporters by value. See Petra Berkhout, *Food Economic Report 2018 of the Netherlands; Summary* (Wageningen Economic Research, 2019).

41. Frank Viviano, "This Tiny Country Feeds the World: Agricultural Giant Holland Is Changing the Way We Farm," *National Geographic* 232, no. 3 (September 2017): 82–94.

42. Lenora Ditzler, "Pixel Farming," in *Countryside, A Report*, ed. Julius Wiedemann (Taschen, 2020), 317. See also Lenora Ditzler et al., "Redefining the Field to Mobilize Three-Dimensional Diversity and Ecosystem Services on the Arable Farm," *European Journal of Agronomy* 122 (2021): 1–15.

43. Lenora Ditzler and Clemens Driessen, "Automating Agroecology: How to Design a Farming Robot Without a Monocultural Mindset?" *Journal of Agricultural and Environmental Ethics* 35, no. 2 (2022), https://doi.org/10.1007/s10806-021-09876-x.

44. Dominique Cuppers, interview with Michiel Scheffer, "Dutch Province to Become Silicon Valley of Agrifood," *F&A Next*, March 11, 2019, https://www.fanext.com/dutch-province-to-become-silicon-valley-of-agrifood.

45. Ditzler, "Pixel Farming," 318.

46. Mads Borup et al., "The Sociology of Expectations in Science and Technology," *Technology Analysis & Strategic Management* 18, nos. 3–4 (2006): 285–298; Harro van Lente et al., "Comparing Technological Hype Cycles: Towards a Theory," *Technological Forecasting and Social Change* 80, no. 8 (October 2013): 1615–1628.

47. John Asafu-Adjaye et al., *An Ecomodernist Manifesto* (Breakthrough Institute, 2015), http://www.ecomodernism.org/manifesto-english; Jonathan Symons, *Ecomodernism: Technology, Politics and The Climate Crisis* (Polity, 2019). As the manifesto argues, for decoupling to happen, a "sustained commitment to technological progress" is required, with "the active, assertive, and aggressive participation of private sector entrepreneurs, markets, civil society, and the state" (29–30).

48. Erik Gómez-Baggethun, "More Is More: Scaling Political Ecology within Limits to Growth," *Political Geography* 76 (2020): 4. A similar argument is made in Erik Gómez-Baggethun, "Sustainable Development," in *Pluriverse: A Post-Development Dictionary*, ed. Ashish Kothari et al. (Tulika, 2019); Jason Hickel and Georgios Kallis, "Is Green Growth Possible?" *New Political Economy* 25, no. 4 (2020): 469–486; and Tim Jackson, *Prosperity Without Growth: Foundations for the Economy of Tomorrow* (Earthscan, 2017).

49. Rem Koolhaas and AMO, *Countryside, The Future*, Guggenheim Museum, New York, February 20, 2020–February 15, 2021. The two images also appear on the front and back cover of the accompanying publication, Julius Wiedemann, ed., *Countryside, A Report*.

50. For critiques of development thinking, see, for example, Majid Rahnema with Victoria Bawtree, eds., *The Post-Development Reader* (Zed, 1997); Escobar, *Encountering Development*; Arturo Escobar, "Imagining a Postdevelopment Era? Critical Thought, Development, and Social Movements," *Social Text* 31, no. 32 (1992): 20–56.

51. Tom Ravenscroft, "AMO and Volkswagen Propose Electric Tractor Design for Sub-Saharan Africa," *Dezeen*, March 2, 2020, https://www.dezeen.com/2020/03/02/amo-volkswagen-electric-tractor-africa.

52. As the authors of a technical paper on another, unaffiliated electric-powered tractor project note, the collaboration between AMO and VW is likely to lead to a product far beyond the price point of what a farmer in a sub-Saharan country such as Malawi could afford. See Toby Williams et al., "Micro-Electric Tractors for Deep Bed Farming and Sustainable Micro-Grid Electricity," *Frontiers in Energy Research* 12 (2024), https://doi.org/10.3389/fenrg.2024.1384568.

53. Such innovation is not always the product of capitalist processes. See, for example, Mariana Mazzucato, *The Entrepreneurial State: Debunking Public vs. Private Sector Myths* (Penguin, 2024).

54. The term "weak signals" was coined by H. Igor Ansoff, in "Managing Strategic Surprise by Response to Weak Signals," *California Management Review* 18, no. 2 (Winter 1975): 21–33.

55. Rem Koolhaas, "*Countryside, The Future*—Teaser Video Transcript," https://www.guggenheim.org/wp-content/uploads/2020/01/guggenheim-countryside-teaser-video-transcript.pdf.

56. Tony Fry, *Defuturing: A New Design Philosophy* (Bloomsbury, 2020). As Fry puts it, "Fundamentally, we act to defuture because we do not understand how the values, knowledge, worlds and things we create go on designing after we have designed and made them" (10).

57. *Buen vivir* was written into the Ecuadorian constitution in 2008, and the Bolivian constitution in 2009. *Sumak kawsay* has an analogous concept in Aymara as *Suma Qamaña*, although the two cannot be seen as identical. As Chuji, Grimaldo, and Gudynas point out, "It is incorrect to say that *buen vivir* is exclusively an indigenous proposition, as [it is] to say that it implies returning to a pre-colonial condition, although those contributions are essential to its construction." See Mónica Chuji et al., "Buen Vivir," in *Pluriverse: A Post-Development Dictionary*, ed. Ashish Kothari et al. (Tulika, 2019), 112.

58. De Zaldívar provides a critical discussion of three strands in the discourse on buen vivir and sumak kawsay. At the two extremes, he cites René Ramírez as one of the key figures in the state-led use of the term, and Atawallpa Oviedo as a vociferous defender of an Indigenous conception of sumak kawsay. The work of Alberto Acosta represents a third strand, which converges with Indigenous perspectives, while attempting to evade charges of essentialism and romanticism that are sometimes directed at the second camp. See Víctor Bretón Solo de Zaldívar, "Three Divergent Understandings of Buen Vivir in the Ecuador of the Citizens' Revolution," *Latin American and Caribbean Ethnic Studies* 12, no. 2 (2017): 188–198.

59. Eduardo Gudynas, "Buen Vivir: Today's Tomorrow," *Development* 54, no. 4 (2011): 445.

60. Carlos Viteri Gualinga, "Visión indígena del desarrollo en la amazonía," *Polis* 3 (2000), http://polis.revues.org/7678, quoted in Zaldívar, "Three Divergent Understandings of Buen Vivir," 193.

61. For a discussion of counter-modern ways of living within diverse Indigenous cultures, see Carolyn Smith-Morris and Cesar Abadia, eds., *Countering Modernity: Communal and Cooperative Models from Indigenous Peoples* (Routledge, 2024).

62. Chuji et al., "Buen Vivir," 112. For further discussion of these temporalities, where Indigenous cosmovisions reveal how "the past-future is contained in the present" (96), see Silvia Rivera Cusicanqui, "*Ch'ixinakax utxiwa*: A Reflection on the Practices and Discourses of Decolonization," *South Atlantic Quarterly* 111, no. 1 (2012): 95–109.

63. Giacomo D'Alisa et al., eds., *Degrowth: A Vocabulary* (Routledge, 2015). See also Arturo Escobar, "Degrowth, Postdevelopment, and Transitions: A Preliminary Conversation," *Sustainability Science* 10, no. 3 (2015): 451–462.

64. Gudynas writes that "unlike degrowth, Buen Vivir, due to its intercultural perspective, follows more ambitious objectives placed in changing present-day cosmovisions of humans, society and Nature." See Eduardo Gudynas, "Buen Vivir," in *Degrowth: A Vocabulary*, ed. Giacomo D'Alisa et al. (Routledge, 2015), 204. The Eurocentrism of degrowth is evident from a glance at a map on the degrowth network website, which (at the time of writing) indicates that only four groups in South America had contributed to the survey. See https://map.degrowth.net. See also Adrián E. Beling et al., "Discursive Synergies for a 'Great Transformation' Towards Sustainability: Pragmatic Contributions to a Necessary Dialogue Between Human Development, Degrowth, and Buen Vivir," *Ecological Economics* 144 (2018): 304–313; Ashish Kothari et al., "Buen Vivir, Degrowth and Ecological Swaraj: Alternatives to Sustainable Development and the Green Economy," *Development* 57 (2014): 362–375.

65. On this more state-centered anticipatory politics, see Ben Anderson, "Preemption, Precaution, Preparedness: Anticipatory Action and Future Geographies," *Progress in Human Geography* 34, no. 6 (2010): 777–798.

66. One of several recent legal judgements on the "Rights of Nature" in Ecuador was issued in 2021, when the Constitutional Court of the country concluded that these rights had been violated by mining and exploration activities in the Los Cedros cloud forest carried out by Ecuadorian company Enami and its Canadian partner Cornerstone. These actors were ordered not only to withdraw from Los Cedros but also to remediate the damage they had caused through restoration activities. See Katie Surma, "Ecuador's High Court Affirms Constitutional Protections for the Rights of Nature in a Landmark Decision," *Inside Climate News*, December 3, 2021, https://insideclimatenews.org/news/03122021/ecuador-rights-of-nature.

67. Craig Jeffrey and Jane Dyson, "Geographies of the Future: Prefigurative Politics," *Progress in Human Geography* 45, no. 4 (2021): 644–645. For another definition, see Carl Boggs, "Marxism, Prefigurative Communism, and the Problem of Workers' Control," *Radical America* 11, no. 6 (1977): 101. "The embodiment, within the ongoing political practice of a movement, of those forms of social relations, decision-making, culture and human experience that are the ultimate goal."

68. Simon Springer, "Human Geography Without Hierarchy," *Progress in Human Geography* 38, no. 3 (2014): 402–419; Rachel Silvey, "Borders, Embodiment, and Mobility: Feminist Migration Studies in Geography," in *A Companion to Feminist*

Geography, ed. Lise Nelson and Joni Seager (Wiley, 2005); Robyn Longhurst and Lynda Johnston, "Bodies, Gender, Place and Culture: 21 Years On," *Gender, Place & Culture* 21, no. 3 (2014): 267–278.

69. Peter Kraftl, *Geographies of Alternative Education: Diverse Learning Spaces for Children and Young People* (Polity, 2013); Emily Brissette, "Prefiguring the Realm of Freedom at Occupy Oakland," *Rethinking Marxism* 25, no. 2 (2013): 218–227; Sam Halvorsen, "Spatial Dialectics and the Geography of Social Movements: The Case of Occupy London," *Transactions of the Institute of British Geographers* 42, no. 3 (2017): 45–57; Anja Habersang, "Utopia, Future Imaginations and Prefigurative Politics in the Indigenous Women's Movement in Argentina," *Social Movement Studies* 23, no. 4 (2022): 479–494; Miriam Meissner, "Towards a Cultural Politics of Degrowth: Prefiguration, Popularization and Pressure," *Journal of Political Ecology* 28, no. 1 (2021): 511–532.

70. Galtung, "On Future Research and Its Role in the World," 113; Céline Granjou et al., "Editorial to the Special Issue," in "Environmental Justice Futures," ed. Céline Granjou et al. (special issue), *Futures* 116 (2025): article 103518; Fanny Lajarthe and Lydie Laigle, "Bringing the Future Back to the Present: The Role of Prefiguration in European Climate Justice Activism," *Futures* 160 (2024): article 103384.

71. Uri Gordon, "Prefigurative Politics between Ethical Practice and Absent Promise," *Political Studies* 66, no. 2 (2018): 521–537, although Jeffrey and Dyson (see note 67 of this chapter) contest this conceptualization of prefigurative politics.

72. Benjamin Franks, *Rebel Alliances: The Means and Ends of Contemporary British Anarchisms* (AK, 2006), 119 and 134.

73. Jeffrey and Dyson, "Geographies of the Future," 646, although the authors take issue with the idea that prefigurative political struggles necessarily need to have clearly defined ends.

74. The artist Andrea Fraser makes a similar point regarding enactment in an effort to find an alternative to the term "performative," which she views as debased in discourse about contemporary art. For her, the term "enactment" "allows us to step back from the opposition between doing, acting, or performing on the one hand, and saying or representing on the other, by framing a focus on what we are doing within or beyond—and often in contrast—to what we are saying." See Andrea Fraser, "Performance or Enactment," in *Performing the Sentence: Research and Teaching in Performative Fine Arts*, ed. Carola Dertnig and Felicitas Thun-Hohenstein (Akademie der bildenden Künst and Sternberg Press, 2014), 126.

75. Kathrin Böhm, in conversation with the author, September 18, 2021, The Showroom Gallery, London.

76. The Centre for Plausible Economies (Kathrin Böhm and Kuba Szreder), *Icebergian Economies of Contemporary Art* (Pyrammmida, 2020), 20. See also Stephen Wright, *Toward a Lexicon of Usership* (Van Abbemuseum, 2013), for a similar line of thinking.

77. It is worth noting that Böhm is not alone in using the composting as a metaphor, which has become something of a leitmotif in environmental humanities and science and technology studies in recent years. See Donna Haraway, *Staying with the Trouble: Making Kin in the Chthulucene* (Duke University Press, 2016); Bruno Latour, *Down to Earth: Politics in the New Climatic Regime*, trans. Catherine Porter (Polity, 2018), 86; and Jennifer Mae Hamilton and Astrida Neimanis, "Composting Feminisms and Environmental Humanities," *Environmental Humanities* 10, no. 2 (2018): 501–527.

78. *Potential Agrarianisms: Will There Still Be Sugar After the Rebellion?*, curated by Maja and Reuben Fowkes, Kunsthalle Bratislava, August 21–October 31, 2021.

79. For a historical account of hop picking, see Gilda O'Neill, *Pull No More Bines: Hop Picking—Memories of a Vanished Way of Life* (Women's Press, 1990).

80. Company Drinks promotional video, edited by Sue Giovanni, 2019, https://player.vimeo.com/video/326901966.

81. J. K. Gibson-Graham and Kelly Dombroski, "Introduction: Inventory as Ethical Intervention," in *The Handbook of Diverse Economies*, ed. J. K. Gibson-Graham and Kelly Dombroski (Edward Elgar, 2020), 3 and14.

82. Gibson-Graham draws on Resnick and Wolff's theories of overdetermination. See Stephen Resnick and Richard Wolff, "Marxist Epistemology: The Critique of Economic Determinism," *Social Text* 6 (Autumn 1982): 31–72. This in turn draws upon the work of Louis Althusser, who coined the term. See Stephen Resnick and Richard Wolff, "On Overdetermination and Althusser: Our Response to Silverman and Park," *Rethinking Marxism* 25, no. 3 (2013): 341–349.

83. J. K. Gibson-Graham, *A Postcapitalist Politics* (University of Minnesota Press, 2006), 56.

84. Gerrie van Noord et al., eds., *Kathrin Böhm: Art on the Scale of Life* (Sternberg, 2023), 148.

85. For a critique of scale-invariance, see Derek Woods, "Scale Critique for the Anthropocene," *Minnesota Review*, no. 83 (2014): 133–142.

86. Brecht De Smet, "Alterglobalization and the Limits of Prefigurative Politics," in *Collaborative Projects: An Interdisciplinary Study*, ed. Andy Blunden (Brill, 2014); Mike Hodson and Simon Marvin, "Urbanism in the Anthropocene: Ecological Urbanism or Premium Ecological Enclaves?" *City* 14, no. 3 (2010): 298–313; Lucía Argüelles et al., "Power and Privilege in Alternative Civic Practices: Examining Imaginaries of Change and Embedded Rationalities in Community Economies," *Geoforum* 86 (2017): 30–41.

87. Fernando García-Dory, *Bionic Sheep*, 2006–, https://www.fernandogarciadory.info/index.php?/projects/bionic-sheep.

88. The United Nations Development Programme—Programme of Assistance to the Palestinian People (UNDP-PAPP), *Resilience Series: Agriculture in Area C*, March 2017, https://www.undp.org/sites/g/files/zskgke326/files/2022-11/resilience_series_-_agriculture_in_area_c_-_final.pdf.

89. Anthony Downey, ed., *Nida Sinnokrot: Palestine Is Not a Garden* (Sternberg, 2025).

90. While every Documenta attracts widespread media coverage, in 2022, a large proportion of this was dominated by debates around antisemitism, prompted by the inclusion of Taring Padi's *People's Justice* (2002), a public mural featuring antisemitic tropes, but extending to participating artists including the Palestinian group Question of Funding, for their alleged support of the BDS movement, and Archives des Luttes des Femmes en Algérie, for representations of the Israeli military. See Alex Greenberger, "Documenta's Anti-Semitism Controversy, Explained: How a German Art Show Became the Year's Most Contentious Exhibition," *Artnews*, July 22, 2022, https://www.artnews.com/art-news/news/what-is-documenta-15-antisemitism-controversy-1234635001.

91. ruangrupa, Artistic Team, et al., *Documenta Fifteen Handbook* (Hantje Cantz, 2022), 21.

92. As ruangrupa notes on the Documenta website, the kios was inspired by the "Feral Trade" model developed by the artist Kate Rich. See "Working Group: *Lumbung Kios*," documenta fifteen, https://documenta-fifteen.de/en/lumbung-kios, accessed July 1, 2023; and "Statement," Feral Trade, https://feraltrade.org/statement, accessed July 1, 2023.

93. For more information see the DAYRA project website, https://dayra.net.

94. Gibson-Graham also draws inspiration from the "pluriverse" epistemologies referenced above, emerging from Indigenous worldviews.

95. Anderson, "Preemption, Precaution, Preparedness."

96. Ursula K. Le Guin, "A Non-Euclidean View of California as a Cold Place to Be," in *Dreams Must Explain Themselves: The Selected Non-Fiction of Ursula K. Le Guin* (Gollancz, 2018), 110–111.

CONCLUSION

1. Margaret Atwood, "It's Not Climate Change—It's Everything Change," *Medium*, July 27, 2015, https://medium.com/matter/it-s-not-climate-change-it-s-everything-change-8fd9aa671804. See also Amitav Ghosh, *The Great Derangement: Climate Change and the Unthinkable* (Chicago University Press, 2016).

2. Ben Anderson, "Preemption, Precaution, Preparedness: Anticipatory Action and Future Geographies," *Progress in Human Geography* 34, no. 6 (2010): 782.

3. For further reading on lock-in and path dependency, see note 71 in the introduction.

4. Catriona McKinnon, "Sleepwalking into Lock-In? Avoiding Wrongs to Future People in the Governance of Solar Radiation Management Research," *Environmental Politics* 28, no. 3 (2019): 441–459; Seth Baum et al., "Double Catastrophe: Intermittent Stratospheric Geoengineering Induced by Societal Collapse," *Environment Systems & Decisions* 33, no. 1 (2013): 168–180.

5. McKinnon, "Sleepwalking into Lock-In?" 450.

6. Toni Ahlqvist and Martin Rhisiart, "Emerging Pathways for Critical Futures Research: Changing Contexts and Impacts of Social Theory," *Futures* 71 (2015): 91–104.
7. For an example of linguistic lock-in, see the surprisingly recent advent of the word "hello," coined by Thomas Edison as a way of answering the telephone in the 1870s. See Carl Shapiro and Hal R. Varian, *Information Rules* (Harvard Business School Press, 1999), 207–208.
8. S. Beck and Jeroen Oomen, "Imagining the Corridor of Climate Mitigation—What Is at Stake in IPCC's Politics of Anticipation?" *Environmental Science and Policy* 123 (2021): 169–178.
9. Johan Galtung, "On Future Research and Its Role in the World," in *Challenges from the Future: Proceedings of the International Future Research Conference*, vol. 1, compiled by the Japan Society of Futurology (Kodansha, 1970), 114.

INDEX

Page numbers in *italic* indicate figures. Page numbers in the form of 181n19 indicate endnotes.

Publisher contact:
The MIT Press
Massachusetts Institute of Technology
77 Massachusetts Avenue, Cambridge, MA 02139
mitpress.mit.edu

EU Authorised Representative:
Easy Access System Europe, Mustamäe tee 50,
10621 Tallinn, Estonia
gpsr.requests@easproject.com

Printed by Integrated Books International,
United States of America